SHOCK DATA ANALYSIS

ENGINEERING PRINCIPLES AND TECHNIQUES

by

Ronald D. Kelly

and

George Richman

Wexford Press
2008

PREFACE

The mechanical shock field has long presented one of the most perplexing challenges to the engineering profession. The intensity of the input from the shock environment, even though it may exist for but a brief period of time, causes it to be of particular concern. The effects of explosive overpressures on men, equipment, and structures; the effects of earthquakes on structures; the effects of an automobile crash on its occupants; and the effects of dropping delicate equipment on the floor; all of these are examples of the many situations where the shock environment represents a clear hazard.

Because the solution of shock problems is so important to many application areas, we feel privileged to have been given this opportunity to make a contribution toward easing the solution of shock problems. To solve any specific shock problem, a number of technical disciplines are required. These disciplines include data acquisition, data analysis, data evaluation, design, fabrication, and testing. The attention of our efforts in this book is focused solely on the data analysis discipline.

The one idea that we hope will be most firmly conveyed to the reader is our strong conviction about the manner in which a data analysis technique should be selected. We believe that the technique employed to analyze any particular shock problem should be selected on its ability to provide a satisfactory solution. An analysis technique should not be selected just because "it was always used in the past" or because it is convenient to perform. The merits of the technique in providing the desired solution should be the primary selection criterion—tempered by economic and time considerations, naturally.

The manner in which this book should be read is dependent on the reader's background. For the reader new to the field, it is recommended that the chapters be read in numerical sequence. For those readers already familiar with the field, it is recommended that they skip Chapters 2 and 3 on a first reading, as these deal only with the basic mathematical techniques. Chapters 2 and 3 should serve only as a refresher and as a reference for these readers.

This book was prepared under the generous sponsorship of the Shock and Vibration Information Center at the Naval Research Laboratory. We wish to thank Dr. W. W. Mutch and Mr. H. C. Pusey for the opportunity and support required to complete this work.

We would like to acknowledge the helpful comments of many of our associates, and the numerous typists who prepared the manuscript. In particular, we are grateful for the editorial suggestions of Dr. J. S. Bendat and the drafting support of Kazimierz Niemiec.

RONALD D. KELLY
GEORGE RICHMAN

Los Angeles, California

CONTENTS

Chapter 1

INTRODUCTION

This monograph is strictly concerned with the analysis of transients as applied to mechanical systems. Throughout this text, the particular type of transient concerned is called a shock. To properly lay the groundwork for the subjects to be covered in later chapters, it is first necessary to discuss data types in general and then the distinguishing characteristics of shock data.

1.1 Data Types

As a general rule, the most accurate method for determining the properties of a physical phenomenon consists of directly measuring these properties and then carefully analyzing the measurements to determine the underlying relationships. This is the basis of the scientific method. Any other approach requires that at least portions of the phenomenon be modeled, and this modeling typically requires simplifying assumptions. The errors inherent in these assumptions are almost always greater than the errors involved in measuring the characteristics.

Any observed data may be categorized broadly as either *deterministic* or *nondeterministic* in nature. Deterministic data are those which can be described accurately by some explicit mathematical relationship. That is, any future value may be predicted solely from knowledge of the present value.

As an example of this type of data, consider a series of observations of the motion of a point located on the perimeter of a rotating wheel as shown in Fig. 1.1. It may be shown that the vertical component of the motion of this point is described by the mathematical relationship

$$x(t) = r[\sin(\omega t + \Theta) + 1], \qquad (1.1)$$

where r is the radius of the wheel, ω is its angular velocity, and Θ is the initial angular displacement of the point from the horizontal. Data taken from such a system are deterministic because of the explicit relationship defined by Eq. (1.1).

1

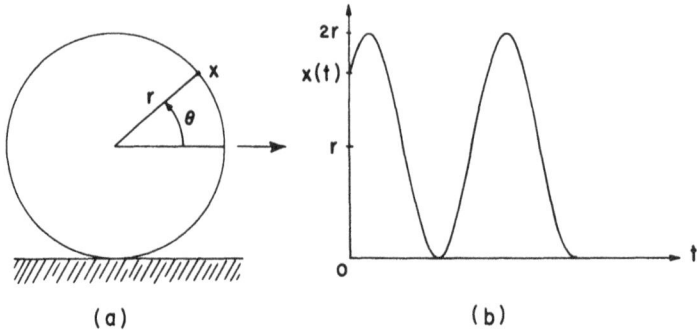

Fig. 1.1 An example of deterministic data; (a) rotating wheel, (b) vertical motion of a point (x) on the wheel.

In practice, many physical phenomena are deterministic in nature. The motion of a satellite in orbit and the temperature of water as heat is applied are two more examples of deterministic data.

Nondeterministic data, as the name implies, are those observations which cannot be described by an explicit mathematical relationship. Instead, the only meaningful statements which can be made about these data are statistical or probabilistic in nature. That is, while it is not possible to predict future values of the data exactly, it is possible to impose bounds upon these future observations with some specified degree of confidence. Such data are often termed *random*, although to use this name properly a rather elaborate set of mathematical conditions must be passed by the data. Practically speaking, the terms *random* and *nondeterministic* are used interchangeably.

As an example of a random phenomenon, consider the acoustic noise of a jet aircraft. This noise will vary randomly because of the complex turbulence caused by the meeting of high-velocity exhaust gases from the jet engine with the atmosphere.

The classification of data into random and deterministic categories is open to question in that observations taken of physical phenomena are almost never truly one or the other. Instead, deterministic data usually contain a random component due to measurement errors, environmental effects, etc., or what may appear to be random data may actually be deterministic data whose causal relationships are so complex as to be unknown at this time. Atmospheric conditions as measured in terms of pressure or temperature are examples of deterministic phenomena which are usually considered random for practical purposes because of the overwhelming difficulties encountered in explicitly defining the causal relationship.

A *time history* may be defined as a series of observations of a phenomenon taken over a specified time interval. A single time history is called a sample function, and the collection of all sample functions for the phenomenon is called a *process*. All random processes may be further classified into stationary and nonstationary categories. In simple terms, a random process is stationary if its moments are invariant with time. For practical applications, this concept of stationarity is frequently restricted to satisfying the condition that only the first two moments, namely the mean and variance, are time-invariant. If these moments change with time, the random process is nonstationary. It should be noted that stationarity requirements as defined here are based upon ensemble averages. However, if the time average of a sample function of a stationary process is identical to the ensemble average taken over the entire process, then the process is said to be *ergodic*. Frequently, in practical applications, ergodicity is assumed if the data are stationary because it is often difficult to acquire more than a few sample functions from the process being studied and, as a result, ensemble averages cannot be taken. As examples of stationary and nonstationary ergodic processes, consider the sample functions shown in Fig. 1.2. The nonstationary character of Fig. 1.2b is visually apparent, but in many instances nonstationarities are difficult to detect.

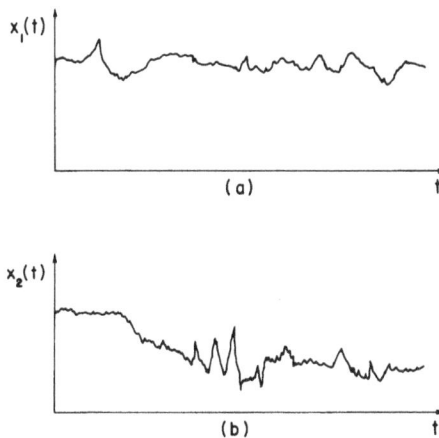

Fig. 1.2. Sample functions (a) stationary process, (b) nonstationary process.

The types of data to be discussed in this monograph may be classified as either deterministic or nonstationary random. When the problem is one of analyzing repeatable shocks, such as might be produced by some shock-testing machines, then the data are deterministic. However, most shocks caused by natural environments, such as those due to the accidental dropping of a component, or the pyrotechnic shock caused when explosive bolts are fired on a space vehicle, are nonstationary random phenomena because of the variability between sample functions.

1.2 Transient Data

The practical definition of what constitutes transient data presents
a problem. Considered from a mathematical point of view, all physical
processes are transient. True sine waves, ramp functions, etc., are
mathematical abstractions rather than observable physical phenomena. For example, even the output of an oscillator will not always
be a sinusoid of the same frequency and amplitude. From an engineering point of view, there are many physical processes that can be considered sufficiently stationary so that the mathematical abstractions
can be used to obtain solutions to practical problems. If the above
oscillator is tuned to a frequency of 1000 Hz and maintains a reasonably
stable frequency and amplitude for an hour, its output is normally
considered to be a pure sinusoid.

It·is much easier to define what is not a transient than to define
what is a transient. A few cases are clearly transients: the single-
pulse category, for example. The acceleration time history of an item
dropped on the floor clearly falls into the transient category. However,
consider the time period during which the amplitude of the above
oscillator is being changed. Is this a transient condition? Again, from
the mathematical viewpoint, it is transient behavior, but from a practical viewpoint there are cases where stationary analysis techniques
can be used. This is desirable when possible because stationary analysis
procedures are generally much simpler than transient analysis procedures. The key to classification usually lies in the rate of change of
the input conditions relative to the system to which the input is applied.
Suppose that the output of the above oscillator is fed to a second order
filter resonant at 100 Hz. Let the amplitude be changed linearly by some
factor A in a time period τ. If τ is 10 min, the output at any time during
this period can be accurately predicted. It will be a sine wave of the
same frequency as the input, and its magnitude will vary linearly with
time by the same factor A. If, on the other hand, the change in level
occurs in 1 msec, transient analysis techniques must be used to predict
the output during the change and shortly thereafter. (For a solution
to this problem see Section 5.2). Thus, the classification of a particular
time history as transient or not depends on how the system of interest
responds to this time history. Are the transient response equations
necessary to adequately describe the response, or can they be ignored
and only the steady state equations be used?

Mechanical shocks are usually of great interest in the design and
operation of physical systems because the instantaneous input levels
are frequently an order of magnitude or more higher than the steady
state inputs. Examples of several transient time histories are shown
in Figs. 1.3 through 1.7. These time histories are from an earthquake,
an impact in a railroad car, a torpedo hitting the water, a pyrotechnic
shock, and a nuclear explosion, respectively.

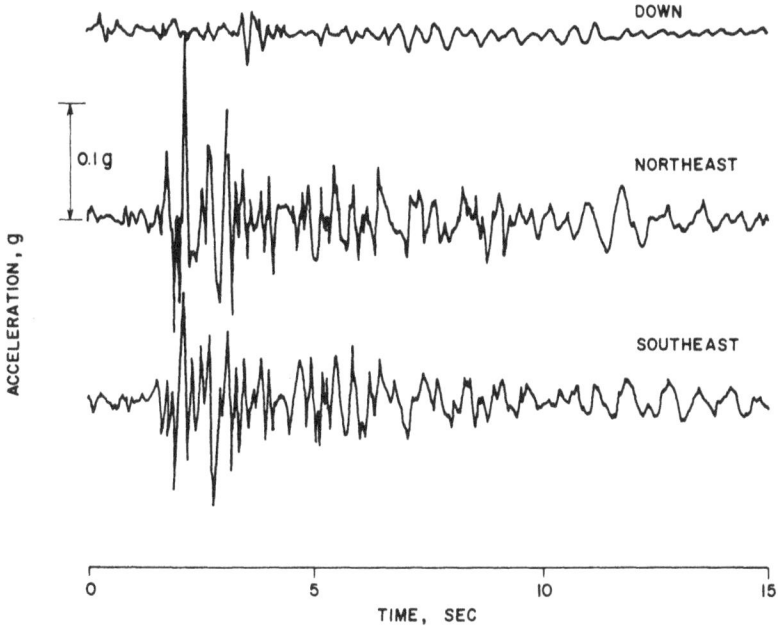

Fig. 1.3. Time history of an earthquake shock. (From Ref. 1,
reprinted by permission of the Seismological Society of America.)

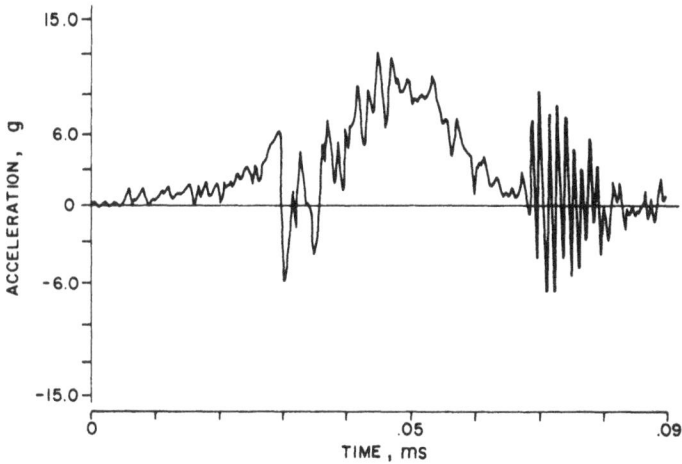

Fig. 1.4. Time history of a railroad car impact (Ref. 2).

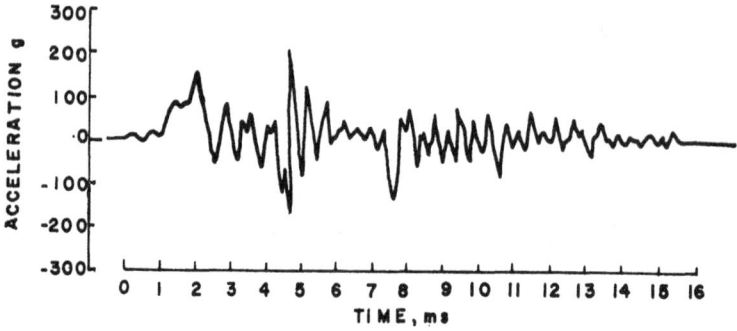

Fig. 1.5.   Shock time history of a torpedo water entry (Ref. 3).

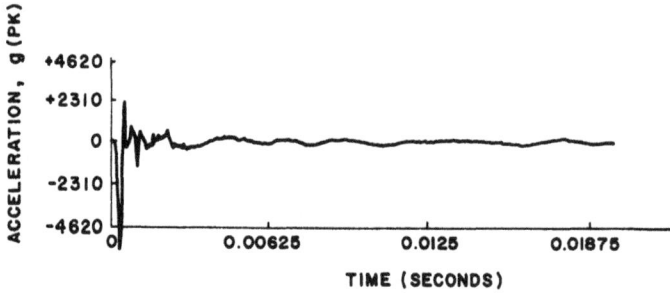

Fig. 1.6.   Time history of a pyrotechnic shock (Ref. 4).

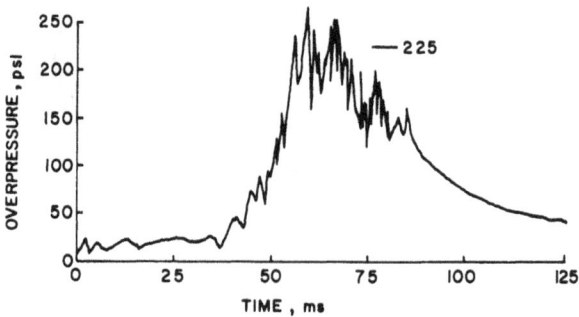

Fig. 1.7.   Shock time history of a
nuclear explosion (Ref. 5).

Note that in all of these, the instantaneous values attain tremendously high levels for that system. Thus, the shock environment is an extremely important one that must be considered in the design of many physical systems.

1.3 Transient Analysis

In the context of this monograph, the underlying common denominator of transient analyses is the determination of the damage potential of a shock upon a physical system. Analyses are concerned either with the design of the system to survive the shock environment or with the attenuation of the shock input to the system by means of packaging or attenuation devices.

"Survival" of a shock excitation can have two entirely different meanings:

● The system exhibits no permanent damage after the shock, or
● The system exhibits no degradation of performance either during or after the shock.

As an example of the first definition, consider a radio which has been accidentally dropped on the floor. Whether the radio plays properly during the drop is of no concern. In fact, even if the tuning and volume controls have to be readjusted after the drop, it is of no concern as long as the radio is not permanently damaged by the fall and plays properly afterward.

As an example of the second definition, consider the launching of a guided missile. If the shock from the ignition of the rocket motor causes even a brief malfunction of the missile guidance or control system, the missile will lose its inertial reference and miss its target. Note that in this case permanent damage to the system is not required in order to fail the second definition of survival.

The design of a system to withstand its shock environment requires the definition of this environment with reasonable accuracy because survival is not the only design factor. For example, weight and size are also frequently important and, unfortunately, usually inversely related to the shock resistance of a system. The strength of a system is usually weight-dependent, while the packaging of the system to reduce shock input is usually size-dependent. However, in many applications, size and weight must be minimized – for example, in space applications where small weight increases require large increases in booster performance. Thus, the system must be able to withstand the shock environment but cannot have a large safety factor because of the weight restriction. This requires an accurate definition of the environment.

In addition to the shocks encountered by a system in its service environment, it may have to undergo testing in the laboratory. These tests will normally include both design verification and proof of workmanship. The former is usually a more severe test where the system is subjected to an environment intended to be at least as severe as the service environment the system is to withstand. Until recently, however, the time histories of the design verification shocks bore little resemblance to those encountered during service. Rather, they were based upon duplicating the damaging effects of the environment. Recently, some effort has been expended on developing the capability to reproduce measured shock time histories as laboratory test excitations.

The specific problem discussed in this monograph is that of analyzing excitation data acquired in time history form. The emphasis is on the reduction of these data to extract those parameters which can be used to solve the engineering problem being studied while rejecting the unimportant parameters that tend to obscure the solution.

Because the failure of a system is directly related to its response to its shock environment, the emphasis in shock analysis has been placed primarily on determining this response. This has been performed either by decomposing the time history into simpler mathematical functions in order to facilitate the computation of the system response, or by determining the failure-related parameters in the response of an estimated model of the system to the shock time history.

Unfortunately, all the mechanisms by which shocks cause failures in physical systems are not well understood. It is commonly assumed that a system can fail either because of a single, extremely high, response amplitude or because of fatigue damage accumulated over many response cycles. The latter failure can occur during a single shock or after exposure to multiple shocks. The first failure mechanism is usually referred to as the single highest peak criterion, and the parameter to be determined from the time history is the peak response of the system. The second mechanism is based on a fatigue failure criterion, and the parameters of importance are the number and amplitude of the stress reversals indicated by the relative maxima and minima of the system response.

While it is possible to determine the required parameters directly from the shock time history under certain simplifying assumptions, a majority of the techniques used to date perform the analysis in the frequency domain. Analysis in this domain is usually termed *spectral* decomposition. The advantages of spectral procedures are manyfold. First of all, physical systems may be modeled most easily and accurately in this domain. Many structural models, for example, consist of a set of simple second order oscillators connected in a manner which simulates the various components of the particular structure. Such a model is shown in Fig. 1.8. Each oscillator has its own oscillatory or resonant frequency which is dependent upon the size, shape, and material of the component it is simulating. Since each oscillator will be excited primarily by energy at its resonant frequency, the knowledge of the frequency content of the shock provides the key to the determination of the system response to the shock. Because of these reasons, spectral decomposition techniques will receive primary consideration in this monograph.

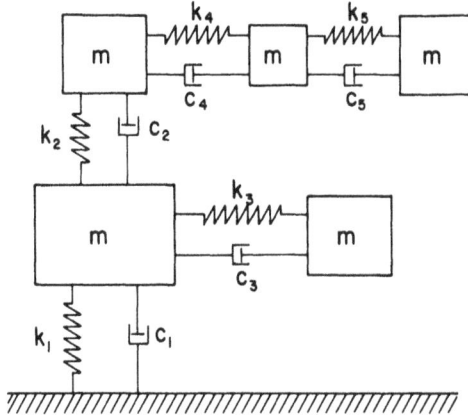

Fig. 1.8. Structural model composed of simple
second order systems.

1.4 Organization of the Monograph

The remaining chapters of this monograph are divided into two
primary subjects. The next three chapters contain basic background
material to introduce the subject of shock analysis. The fundamental
mathematical techniques required to solve transient problems are
discussed in Chapter 2; these include differential equations, Fourier
transforms and Laplace transforms. In Chapter 3, techniques for
determining the response of linear systems to transient inputs are
described. In Chapter 4, spectral decomposition is defined and discussed.
Particular emphasis is placed on the two forms of spectral decomposition most commonly used to solve shock problems – Fourier spectra
and shock spectra. These two decompositions are defined and compared
from an applications point of view.

The last three chapters contain discussions of advanced special
analog and digital material on techniques for analyzing shock data.
Some of this material is not available elsewhere. Chapter 5 describes
analog techniques for computing Fourier and shock spectra. It also
presents practical formulas for estimating error magnitudes associated
with these analyses. Chapter 6 presents similar material on digital
techniques for analyzing shock data. Chapter 7 covers methods other
than the Fourier and shock spectra analysis techniques. A broad range
of techniques, from simple single number analyses up to complicated
random transient analyses, is described.

Chapter 2

BASIC MATHEMATICAL PROCEDURES

2.1 Differential Equations

Physical systems may be analyzed in several ways. The most obvious approach is to determine the differential equations describing the operation of the system and then to solve these equations by some analytic procedure. To attain this goal, the system must be relatively simple, since solutions can be obtained by classical techniques only for limited cases.

Simple systems usually give rise to linear, nth order differential equations. The coefficients of the differential equation describe system parameters which are usually invariant with time. The order of the system is equal to twice the number of degrees of freedom available to the response motion, where each degree of freedom is defined as the ability to move along or about an axis.

An example of the simple physical system is the second order, mechanical system shown in Fig. 2.1.

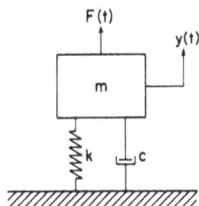

Fig. 2.1. A second order mechanical system.

The system consists of a mass m connected to an immovable foundation by a spring and a dashpot. The spring is linear and has a coefficient k. The viscous damping coefficient of the dashpot is c. If the system is excited by a force $F(t)$ applied to the mass, the differential equation describing the response motion of the mass may be derived by specifying that the sum of all the forces acting on the mass be identically zero. The forces consist of

$$F_k(t) = \text{spring force}$$
$$F_c(t) = \text{damping force}$$

11

$$F_m(t) = \text{inertial force}$$
$$F(t) = \text{exciting force,}$$

and the equation of motion is

$$F_k(t) + F_c(t) + F_m(t) + F(t) \equiv 0. \tag{2.1}$$

The quantity $y(t)$ is defined as the inertial motion of the mass. Then

$$\frac{dy}{dt} = \text{inertial velocity of the mass,}$$

and

$$\frac{d^2y}{dt^2} = \text{inertial acceleration of the mass.}$$

Expressions for the forces in terms of $y(t)$, $\frac{dy}{dt}$, and $\frac{d^2y}{dt^2}$ may be obtained. They are

$$F_k(t) = -ky(t) \tag{2.2}$$

$$F_c(t) = -c\frac{dy}{dt} \tag{2.3}$$

$$F_m(t) = -m\frac{d^2y}{dt^2}. \tag{2.4}$$

Equation (2.1) may now be rewritten as

$$F(t) - ky(t) - c\frac{dy}{dt} - m\frac{d^2y}{dt^2} = 0 \tag{2.5a}$$

or

$$m\frac{d^2y}{dt^2} + c\frac{dy}{dt} + ky(t) = F(t). \tag{2.5b}$$

A closed-form solution for $y(t)$ can now be obtained. The complete solution consists of the sum of the general solution to the homogeneous equation

$$m\frac{d^2y}{dt^2} + c\frac{dy}{dt} + ky(t) = 0 \tag{2.6}$$

and a particular solution of the differential equation itself.

Since the equation is of second order, two solutions exist for the homogeneous equation. The general solution consists of their sum. A solution of the form $y = e^{\lambda t}$ is assumed. Then

$$\frac{dy}{dt} = \lambda e^{\lambda t}, \qquad \frac{d^2y}{dt^2} = \lambda^2 e^{\lambda t}.$$

Equation (2.6) may be rewritten as

$$m\lambda^2 e^{\lambda t} + c\lambda e^{\lambda t} + ke^{\lambda t} = 0. \tag{2.7}$$

Removing the common factor $e^{\lambda t}$, a quadratic equation in λ is obtained.

$$m\lambda^2 + c\lambda + k = 0. \tag{2.8}$$

The two roots of this equation are determined by means of the quadratic formula

$$\lambda_1 = \frac{-c + \sqrt{c^2 - 4mk}}{2m}, \qquad \lambda_2 = \frac{-c - \sqrt{c^2 - 4mk}}{2m}. \tag{2.9}$$

The general solution to the homogeneous equation is then

$$y(t) = c_1 e^{\lambda_1 t} + c_2 e^{\lambda_2 t}, \tag{2.10}$$

where c_1 and c_2 are constants of integration.

The particular solution to the differential equation may be determined by assuming a complex periodic form for the exciting force, i.e.,

$$F(t) = a e^{j\omega t}, \tag{2.11}$$

and then predicting a response motion of a similar nature,

$$y(t) = b e^{j\omega t}. \tag{2.12}$$

Equation (2.5b) may be rewritten as

$$m j^2 \omega^2 b e^{j\omega t} + c j\omega b e^{j\omega t} + k b e^{j\omega t} = a e^{j\omega t}. \tag{2.13a}$$

After removing the common factor $e^{j\omega t}$,

$$-m\omega^2 b + jc\omega b + kb = a \tag{2.13b}$$

or

$$b = \frac{a}{-m\omega^2 + jc\omega + k}. \tag{2.14}$$

Equation (2.14) may also be written as

$$b = a\alpha e^{-j\omega\delta}, \tag{2.15}$$

where

$$\alpha^2 = \frac{1}{(k - m\omega^2)^2 + c^2\omega^2} \tag{2.16}$$

$$\delta = \frac{1}{\omega} \sin^{-1} c\omega\alpha. \tag{2.17}$$

The complete solution is then

$$y(t) = c_1 e^{\lambda_1 t} + c_2 e^{\lambda_2 t} + a\alpha e^{j\omega(t-\delta)}. \tag{2.18}$$

Solving low order differential equations by classical procedures is not difficult. However, as the system becomes more complex, more degrees of freedom are required to describe its motion. This gives rise to higher order equations whose solutions require considerable effort. As a result, other procedures are usually employed.

2.2 Operational Calculus

The solution of differential equations by classical techniques can be quite laborious and time-consuming. In this section several more convenient methods for solving certain types of differential equations are discussed. All of these discussions are restricted to the linear, constant-coefficient class of equations. The first of these operational calculus techniques was developed by Heaviside [6]. All of the other solution techniques that will be covered in this section are variations of this fundamental technique.

In the Heaviside method, all of the derivatives in the differential equations are replaced by a linear operator p. As an example, consider the following equation of a system with an input $x(t)$ and an output $y(t)$:

$$a_1 \frac{d^m x}{dt^m} + a_2 \frac{d^{m-1}x}{dt^{m-1}} + \ldots + a_m \frac{dx}{dt} + a_{m+1}x$$

$$= b_1 \frac{d^n y}{dt^n} + b_2 \frac{d^{n-1}y}{dt} + \ldots + b_n \frac{dy}{dt} + b_{n+1}y. \qquad (2.19)$$

This is rewritten as

$$A[p]x(t) = B[p]y(t), \qquad (2.20)$$

where

$$A[p] = a_1 p^m + a_2 p^{m-1} + \ldots + a_m p + a_{m+1}$$
$$B[p] = b_1 p^n + b_2 p^{n-1} + \ldots + b_n p + b_{n+1}$$

$$p = \frac{d}{dt}, \quad p^2 = \frac{d^2}{dt^2}, \quad p^3 = \frac{d^3}{dt^3}, \quad \text{etc.}$$

If $x(t)$ and $y(t)$ and all of their derivatives are equal to zero at time zero then the solution for the system output at any time greater than zero for some input $x(t)$ is given by the following equation:

$$y(t) = \frac{A[p]}{B[p]} x(t). \qquad (2.21)$$

This solution for $n > m$ is calculated by the Heaviside expansion theorem. For example, let the input be complex periodic and equal to $x_{max}e^{j\omega t}$. Then

$$y(t) = x_{max} \frac{A(j\omega)}{B(j\omega)} + \sum_{r=1}^{n} \frac{A(p_r)e^{p_r t}}{(p_r - j\omega)B'(p_r)}, \qquad (2.22)$$

where

$$p_r = \text{the roots of the equation, } B[p] = 0,*$$

---

*These roots are not repeated and are not zero.

$B'(p_r) =$ the derivative of the polynomial $B[p]$ evaluated at the rth root.

If the input is $x_{max} \cos \omega t$, the real part of Eq. (2.22) is the solution and, similarly, if the input is $x_{max} \sin \omega t$, the imaginary part of this equation is the solution. If the input is x_{max} times a unit step function, the solution is given by Eq. (2.22) with all the ω's set equal to zero.

Integral Transforms

A more convenient approach to the solution of this category of equations is through the use of linear integral transforms. A linear integral transform of a function $x(t)$ is given by

$$T[x(t)] = \int_a^b K(t, u)x(t)dt, \tag{2.23}$$

where

$T[x(t)] =$ an integral transform of $x(t)$,

$K(t, u) =$ a kernel; some particular function of both t and u.

The linearity statement on the transform means that the transform of a sum of two functions is equal to the sum of the two separate transforms and that the transform of a function multiplied by a constant is equal to the same constant multiplied by the transform of the function. Expressed in equation form, this is

$$T[c_1 x_1(t) + c_2 x_2(t)] = c_1 T[x_1(t)] + c_2 T[x_2(t)], \tag{2.24}$$

where c_1 and c_2 are constants. The great value of linear integral transformations is that with certain kernels the transforms of many forms of ordinary differential equations reduce to algebraic equations. This reduces the problem of solving the differential equation to one of determining the roots of an algebraic equation and then taking an inverse transform of this solution.

2.3 Fourier Transforms

If in Eq. (2.23), the kernel is made equal to $e^{-j2\pi ft}$, the lower limit a is set to $-\infty$, and the upper limit b is set to $+\infty$, then the result is known as the Fourier transform

$$F[x(t)] = X(f) = \int_{-\infty}^{\infty} e^{-j2\pi ft}x(t)dt, \tag{2.25}$$

where $F[x(t)] = X(f) =$ the Fourier transform of $x(t)$. Thus the function has been transformed from the time domain to the frequency domain.

Fourier transformation techniques are applied to the solution of differential equations by rewriting Eq. (2.19) as

$$A[j2\pi f]X(f) = B[j2\pi f]Y(f),\qquad(2.26)$$

where

$$A[j2\pi f] = a_1(j2\pi f)^m + a_2(j2\pi f)^{m-1} + \ldots + a_m(j2\pi f) + a_{m+1}$$
$$B[j2\pi f] = b_1(j2\pi f)^n + b_2(j2\pi f)^{n-1} + \ldots + b_n(j2\pi f) + b_{n+1}.$$

The transform of $y(t)$ as a function of the transform of $x(t)$ is found as follows:

$$Y(f) = \frac{A[j2\pi f]}{B[j2\pi f]}X(f).\qquad(2.27)$$

To obtain the solution in the time domain, an integral transformation is made of Eq. (2.27). In this transformation from the frequency to the time domain, the kernel is $e^{j2\pi ft}$. Notice that this kernel differs from that of the Fourier transform, Eq. (2.25), only in the sign of the power of the exponential. This transformation is known as the *inverse* Fourier transformation:

$$F^{-1}[Y(f)] = y(t) = \int_{-\infty}^{\infty} e^{j2\pi ft}Y(f)\,df,\qquad(2.28)$$

where $F^{-1}[Y(f)] = $ the inverse Fourier transform.

As an example of a Fourier transform, consider the time function shown in Fig. 2.2.

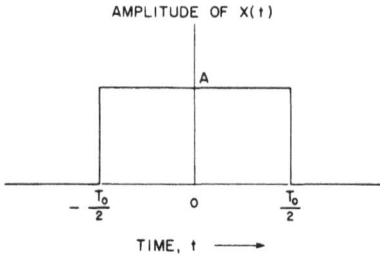

AMPLITUDE OF X(t)

Fig. 2.2.   A boxcar time function.

This time function can be described by

$$x(t) = \begin{cases} A & -\dfrac{T_0}{2} \leq t \leq \dfrac{T_0}{2} \\ 0 & \text{elsewhere.} \end{cases}$$

The Fourier transform of this time function is

$$X(f) = \int_{-T_0/2}^{+T_0/2} Ae^{-j2\pi ft}\,dt = \frac{-A}{j2\pi f}\left(e^{-j2\pi f\frac{T_0}{2}} - e^{j2\pi f\frac{T_0}{2}}\right).$$

By use of Euler's formula, $\sin \alpha = (e^{j\alpha} - e^{-j\alpha})/(2j)$, this equation can be rearranged into the more convenient form of

$$X(f) = (AT_0) \left[\frac{\sin (\pi f T_0)}{\pi f T_0} \right].$$ (2.29)

The magnitude of this $(\sin x)/x$ function is shown in Fig. 2.3. Since the transform is entirely real, the phase factor $[\tan^{-1} (Im/Re)]$ will be zero for all frequencies.

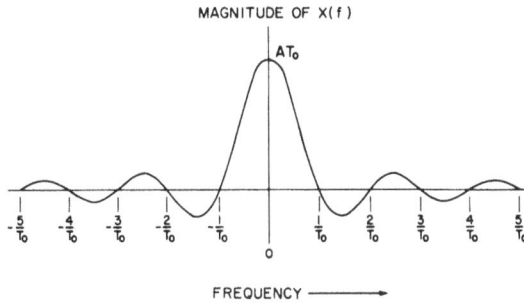

MAGNITUDE OF X(f)

Fig. 2.3. A boxcar frequency function.

As a second example, consider the time function shown in Fig. 2.4.

x(t)

Fig. 2.4. Time function, initial-peak sawtooth.

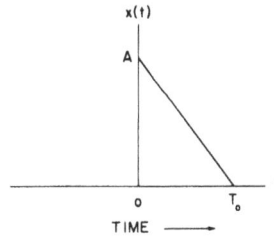

$$x(t) = \begin{cases} A[1 - (t/T_0)] & 0 \leq t \leq T_0 \\ 0 & \text{elsewhere.} \end{cases}$$ (2.30)

The Fourier transform is found as follows:

$$X(f) = \int_0^{T_0} A \left(1 - \frac{t}{T_0}\right) e^{-j2\pi ft} dt$$ (2.31)

$$= \frac{jAT_0}{2\pi f T_0} \left[\left(\frac{\sin \pi f T_0}{\pi f T_0}\right) (e^{-j\pi f T_0}) - 1 \right].$$

The magnitude of this function is shown in Fig. 2.5, and the phase factor in Fig. 2.6.

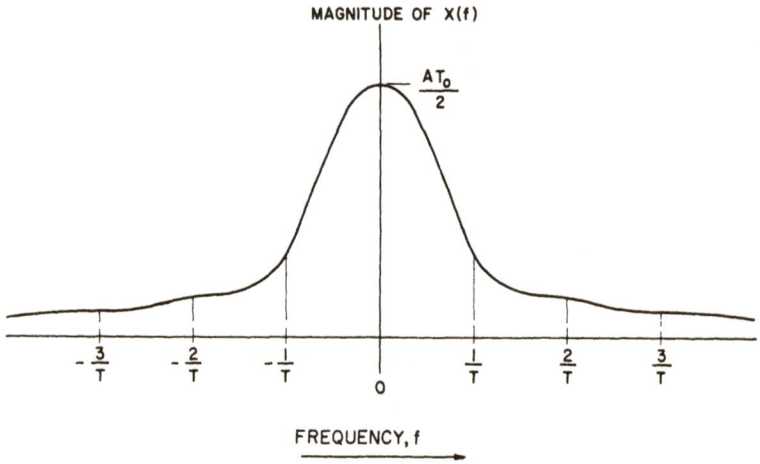

Fig. 2.5.   Frequency function, initial-peak sawtooth.

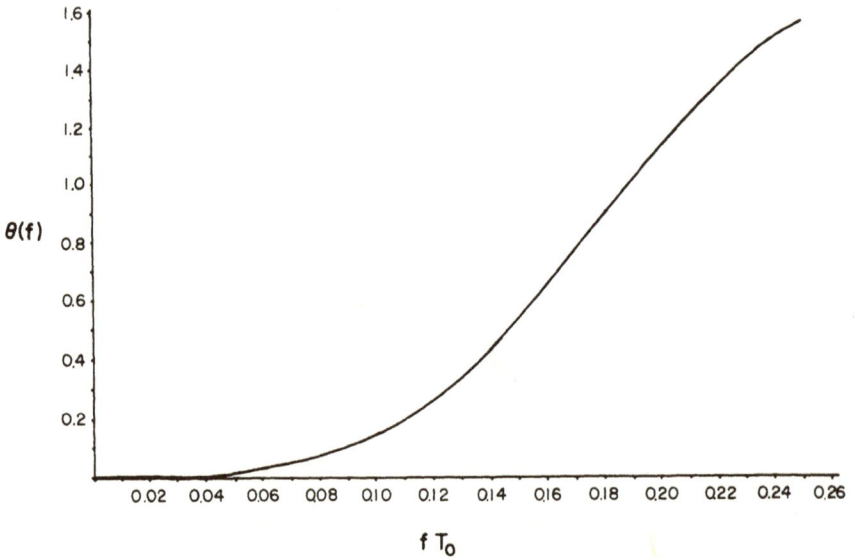

Fig. 2.6.   Phase factor, initial-peak sawtooth.

Now consider the inverse transform of the above two frequency functions. First,

$$x(t) = \int_{-\infty}^{\infty} (AT_0) \left[\frac{\sin (\pi f T_0)}{(\pi f T_0)} \right] e^{j2\pi ft} df. \tag{2.32}$$

Since $e^{jx} = \cos x + j \sin x$, the equation can be rearranged;

$$x(t) = \frac{A}{\pi} \left(\int_{-\infty}^{\infty} \frac{\sin \pi f T_0}{f} \cos 2\pi ft df + j \int_{-\infty}^{\infty} \frac{\sin \pi f T_0}{f} \sin 2\pi ft df \right). \tag{2.33}$$

Since the first term of the above equation is an even function $[f(-x) = f(x)]$ and the second term is an odd function $[f(-x) = -f(x)]$, further simplification is possible and yields

$$x(t) = \frac{2A}{\pi} \int_0^{\infty} \frac{\sin \pi f T_0}{f} \cos 2\pi ft df. \tag{2.34}$$

One further manipulation is required. Let $u = |t|$. Then, since $\cos |x| = \cos x$, the equation becomes

$$x(t) = \frac{2A}{\pi} \int_0^{\infty} \frac{\sin \pi f T_0}{f} \cos 2\pi fu \, dt$$

$$= \frac{2A}{\pi} \left[\begin{array}{ll} \frac{\pi}{2} & u < T_0/2 \\ 0 & u > T_0/2 \end{array} \right] \tag{2.35}$$

or

$$x(t) = \begin{cases} 2A & |t| < T_0/2, \text{ or } -T_0/2 < t < T_0/2 \\ 0 & |t| > T_0/2. \end{cases} \tag{2.36}$$

Thus, the original boxcar time function shown in Fig. 2.2 is obtained.

The inverse transform of the frequency function for the second example is calculated as

$$x(t) = \int_{-\infty}^{\infty} \left\{ \frac{jAT_0}{2\pi f T_0} \left[\left(\frac{\sin \pi f T_0}{\pi f T_0} \right) (e^{-j\pi f T_0}) - 1 \right] \right\} e^{j2\pi ft} df. \tag{2.37}$$

This is expanded to

$$x(t) = \frac{AT_0}{2(\pi T_0)^2} \int_{-\infty}^{\infty} \left(j \frac{\sin \pi T_0 f \cos \pi T_0 f \cos 2\pi tf}{f^2} + \frac{\sin^2 \pi T_0 f \cos 2\pi tf}{f^2} \right.$$

$$- j \frac{\pi T_0 \cos 2\pi tf}{f} - \frac{\sin \pi T_0 f \cos \pi T_0 f \sin 2\pi tf}{f^2} \tag{2.38}$$

$$\left. + j \frac{\sin^2 \pi T_0 f \sin 2\pi tf}{f^2} + \frac{\pi T_0 \sin 2\pi tf}{f} \right) df.$$

By dropping all the odd terms, doubling the value of the even terms, and integrating these over only half the range, the equation simplifies to the following:

$$x(t) = \frac{AT_0}{(\pi T_0)^2} \int_{-\infty}^{\infty} \left(\frac{\sin^2 \pi T_0 f \cos 2\pi t f}{f^2} - \frac{\sin 2\pi T_0 f \sin 2\pi t f}{2f^2} \right.$$

$$\left. + \frac{\pi T_0 \sin 2\pi t f}{f} \right) df. \tag{2.39}$$

The solution for this integral can be found in Ref. 7. The complete solution (inverse transform) is

$$x(t) = \begin{cases} A\left(1 - \dfrac{t}{T_0}\right) & 0 < t < T_0 \\ 0 & T_0 < t, \end{cases} \tag{2.40}$$

which is the original time function.

In Ref. 8, p. 93, an analogy is drawn between a table of logarithms and a table of integral transforms. The purpose of the logarithmic type of transform is to simplify the arithmetical operations of multiplication and division. Thus the primary properties of this transformation are the following:

$$\tau[xy] = \tau[x] + \tau[y]$$
$$\tau[x/y] = \tau[x] - \tau[y]$$
$$\tau[x^n] = n\tau[x],$$

where $\tau[\ ] = $ the logarithmic transform.

A simple table of logarithmic transform pairs is shown in Table 2.1.

TABLE 2.1.   SIMPLE TABLE OF LOGARITHMS

Original Number, x	Logarithmic Transform, $\tau[x]$
1	0
10	1
100	2

Suppose that the multiplication $10 \cdot 10$ is to be performed;

$$\tau[10 \cdot 10] = \tau[10] + \tau[10] = 1 + 1 = 2. \tag{2.41}$$

To get back to the numerical value, an inverse transformation is made simply by using the table;

$$\tau^{-1}[2] = 100. \tag{2.42}$$

(Obviously, these simple multiplications can be carried out faster in a direct manner than by transformation. However, as the functions become more complicated, the transform solution becomes much faster than the direct solution. The same is true of integral transformations.)

From the two examples in this section, a simple Fourier transform table can be constructed (see Table 2.2). Hence, any time that the solution to the differential equations in the frequency domain is of the $(\sin x)/x$ form, the solution in the time domain can be found from the Fourier transform tables to be a boxcar function in the time domain. More extensive tables of Fourier transforms can be found in Refs. 9 and 10.

TABLE 2.2.   SIMPLE TABLE OF FOURIER TRANSFORMS

$x(t)$, *Inverse Transform*	$X(f)$, *Direct Transform*
$A, \quad -\dfrac{T_0}{2} \leqslant t \leqslant +\dfrac{T_0}{2}$ $0, \quad$ elsewhere	$AT_0 \left[\dfrac{\sin \pi f T_0}{\pi f T_0} \right]$
$A\left(1 - \dfrac{t}{T_0}\right), \quad 0 \leqslant t \leqslant T_0$ $0, \quad$ elsewhere	$\dfrac{jAT_0}{2\pi f T_0} \left[\left(\dfrac{\sin \pi f T_0}{\pi f T_0} \right) \left(e^{-j\pi f T_0} \right) - 1 \right]$

Alternate Forms of the Fourier Transform

Fourier transforms are commonly defined in a number of forms. When using tables of Fourier transform pairs, care must be exercised to be sure of the exact definition of the Fourier transform tabulated in that particular table. First, there can be a difference in the scale factor and/or argument of the transforms. In this text, the transform has been defined over f to avoid constants in either the direct or inverse transform. This definition is

$$X(f) = \int_{-\infty}^{\infty} x(t) e^{-j2\pi ft} dt$$

and

$$x(t) = \int_{-\infty}^{\infty} X(f) e^{j2\pi ft} df.$$

Other common forms are defined below, where different subscripts are used on the Fourier transforms to contrast the various definitions:

Form 1

$$X_1(\omega) = \int_{-\infty}^{\infty} x(t)e^{-j\omega t}dt$$

$$(\omega = 2\pi f)$$

$$x(t) = \frac{1}{2\pi} \int_{-\infty}^{\infty} X_1(\omega)e^{j\omega t}d\omega$$

Form 2

$$X_2(\omega) = \frac{1}{2\pi} \int_{-\infty}^{\infty} x(t)e^{-j\omega t}dt$$

$$x(t) = \int_{-\infty}^{\infty} X_2(\omega)e^{j\omega t}d\omega$$

Form 3

$$X_3(\omega) = \frac{1}{\sqrt{2\pi}} \int_{-\infty}^{\infty} x(t)e^{-j\omega t}dt$$

$$x(t) = \frac{1}{\sqrt{2\pi}} \int_{-\infty}^{\infty} X_3(\omega)e^{j\omega t}d\omega.$$

Second, since the kernel of the transformation is a complex exponential, it can be expanded by Euler's formula into cosine and sine components;

$$X(f) = \int_{-\infty}^{\infty} x(t) \cos 2\pi ft dt - j\int_{-\infty}^{\infty} x(t) \sin 2\pi ft dt$$

and

$$x(t) = \int_{-\infty}^{\infty} X(f) \cos 2\pi ft df + j\int_{-\infty}^{\infty} X(f) \sin 2\pi ft df.$$

Frequently, tables will list the Fourier cosine and sine transforms separately from the complex exponential Fourier transforms;

$$X_c(f) = \int_{-\infty}^{\infty} x(t) \cos 2\pi ft dt$$

and

$$X_s(f) = \int_{-\infty}^{\infty} x(t) \sin 2\pi ft dt,$$

where

$$X_c(f) = \text{the Fourier cosine transform}$$
$$X_s(f) = \text{the Fourier sine transform.}$$

Also, some tables define one-sided Fourier cosine and sine transforms;

$$X_{c1}(f) = \int_{0}^{\infty} x(t) \cos 2\pi ft dt$$

and

$$X_{s1}(f) = \int_{0}^{\infty} x(t) \sin 2\pi ft dt.$$

If the equations are either even or odd functions, then the latter types of transforms are particularly useful:

For even functions

$$X(f) = 2X_{c1}(f)$$

For odd functions

$$X(f) = -2jX_{s1}(f)$$

In general

$$X(f) = X_{c1}(f) + X_{c1}(-f) - j[X_{s1}(f) - X_{s1}(-f)].$$

Conditions for the Transform to Exist

Up to this point, nothing has been said about restrictions on the function to be transformed in order that its Fourier transform exist. The reason for this is that almost all physical functions, and certainly practical shock time histories, will satisfy these conditions. Generally, it is only with analytical examples that these conditions cannot be met. The formal conditions are known as the Dirichlet's Conditions and are quoted below from Ref. 11.*

"A function $f(x)$ will be said to satisfy Dirichlet's Conditions in an interval (a, b) in which it is defined, when it is subject to one of the two following conditions:

(i) $f(x)$ is bounded in (a, b), and the interval can be broken up into a finite number of open partial intervals, in each of which $f(x)$ is monotonic.

(ii) $f(x)$ has a finite number of points of infinite discontinuities in the interval. When arbitrary small neighborhoods of these points are excluded, $f(x)$ is bounded in the remainder of the interval, and this can be broken up into a finite number of open partial intervals, in each of which $f(x)$ is monotonic. Further,

the infinite integral

$$\int_{-\infty}^{\infty} f(x)\,dx$$

is to be absolutely convergent."

Special Properties of Fourier Transforms

Since Fourier transforms have been investigated extensively, much is known about their special properties. These properties can be used to good advantage in simplifying analyses. Many of these properties are tabulated in Table 2.3.

---

*Reprinted by permission of Dover Publications, Inc.

TABLE 2.3. SPECIAL PROPERTIES OF FOURIER TRANSFORMS

Property	Function, $x(t)$	Fourier Transform, $X(f)$
Linearity	$ax(t)+by(t)$	$aX(f)+bY(f)$
Convolution	$\int_{-\infty}^{\infty} x(t-\tau)y(\tau)d\tau$	$X(f)\cdot Y(f)$
Multiplication	$x(t)\cdot y(t)$	$\int_{-\infty}^{\infty} X(f-\lambda)Y(\lambda)d\lambda$
Derivative	$\dfrac{dx}{dt}$	$j2\pi f X(f)$
Integral	$\int_{-\infty}^{t} x(\tau)d\tau$	$\dfrac{X(f)}{j2\pi f}, f\neq 0$
Derivative in the Frequency Domain	$-j2\pi t x(t)$	$\dfrac{dX}{df}$
Integral in the Frequency Domain	$-\dfrac{x(t)}{j2\pi t}$	$\int_{-\infty}^{f} X(\lambda)d\lambda$
Contraction in the Time Domain	$x(t/a)$	$\|a\|X(af)$
Contraction in the Frequency Domain	$\|b\|x(bt)$	$X(f/b)$
Translation in the Time Domain	$x(t-a)$	$e^{-j2\pi af}X(f)$
Translation in the Frequency Domain	$e^{j2\pi bt}x(t)$	$X(f-b)$
Area Under the Curve	$\int_{-\infty}^{\infty} x(t)dt=X(0)$	$\int_{-\infty}^{\infty} X(f)df=x(0)$
Energy	$\int_{-\infty}^{\infty} \|x(t)\|^2 dt$	$\int_{-\infty}^{\infty} \|X(f)\|^2 df$
Delta Function in the Time Domain	$x(t)=\delta(t)$	$X(f)=1$

TABLE 2.3. SPECIAL PROPERTIES OF FOURIER TRANSFORMS – Con.

Property	Function, $x(t)$	Fourier Transform, $X(f)$
Delta Function in the Frequency Domain	$x(t)=1$	$X(f)=\delta(f)$
Real-Time Function	$x(t)=\text{Real}$	$X(f)=X^*(-f)$, where $*=$complex conjugate. $Re[X(f)]$ $=Re[X(-f)]$ $Im[X(f)]$ $=-Im[X(-f)]$
Even Time Function	$x(t)=x(-t)$	$X(f)=2\int_0^\infty x(t)\cos 2\pi ft\,dt$
Odd Time Function	$x(t)=-x(-t)$	$X(f)=-j2\int_0^\infty x(t)\sin 2\pi ft\,dt$
Complex Conjugate	$y(t)=x^*(t)$	$Y(f)=X^*(-f)$
Parseval's Theorem	$\int_{-\infty}^{\infty} x^*(t)y(t)\,dt=$ $\int_{-\infty}^{\infty} x(t)y(t)\,dt=$ $\int_{-\infty}^{\infty} x(\tau)Y(f)\,d\tau=$	$\int_{-\infty}^{\infty} X^*(f)Y(f)\,df$ $\int_{-\infty}^{\infty} X(-f)Y(f)\,df$ $\int_{-\infty}^{\infty} X(f)y(\tau)\,d\tau$

Finite Fourier Transforms

In the basic definition of the Fourier transform given by Eq. (2.25), the limits of integration are infinite. In practice, a finite representation of the function is always obtained. The physical record has some starting point that is usually designated as time zero and an end point that will be designated as time T. Thus the Fourier transform of a physical data record is written as

$$\hat{X}(f)=\int_0^T x(t)e^{-j2\pi ft}\,dt, \qquad (2.43)$$

where $\hat{X}(f)=$ the estimate of the Fourier transform of the function $x(t)$ based on a physical record of finite duration.

To use the results previously developed in this chapter, the time base will be defined slightly differently. Specifically, time zero will be defined to be exactly in the middle of the record. Thus, the finite Fourier transform is rewritten as

$$\hat{X}(f) = \int_{-T/2}^{T/2} x(t)e^{-j2\pi ft}dt. \qquad (2.44)$$

If the physical function is of finite duration equal to or less than T and is completely contained in the time interval from $-T/2$ to $+T/2$, no distortion of the true Fourier transform will occur. However, if the true physical function is not completely contained in the interval T, the finite transform will be a distorted version of the true Fourier transform. It then becomes necessary to be able to determine the relation between the error in the transform and the time interval T. By knowing this relationship, it is possible to select the proper duration for the required accuracy in the transform when that option is available. Or, when constrained to a fixed time interval, the accuracy of the finite transform can be estimated. The distortion of the true transform can be evaluated as follows.

The finite transform can be written as the infinite transform of a product of functions where one of the functions is the boxcar function $y(t)$ between $-T/2$ and $+T/2$;

$$\hat{X}(f) = \int_{-\infty}^{\infty} x(t)y(t)e^{-j2\pi ft}dt, \qquad (2.45)$$

where

$$y(t) = \begin{cases} 1 & -T/2 \leqslant t \leqslant +T/2 \\ 0 & \text{elsewhere.} \end{cases} \qquad (2.46)$$

From the table of properties of Fourier transforms, it is noted that the transform of a product of time functions is the convolution of the individual transforms of the pair. Therefore,

$$\hat{X}(f) = \int_{-\infty}^{\infty} X(f-\lambda)Y(\lambda)d\lambda. \qquad (2.47)$$

From Eq. (2.29), $Y(\lambda) = \sin(\pi\lambda T)/(\pi\lambda T)$. It is through this transform that the dependence of the time duration occurs. $X(f)$ is the true Fourier transform and is obtained over an infinite interval; hence, it is independent of T. The transforms for two different record lengths are compared in Fig. 2.7. Note that the side lobes have about the same maximum values, but that the main lobe of the longer duration boxcar function has a greater value and is narrower than the shorter duration function.

The way the finite duration distorts the Fourier transform can be examined by graphically evaluating the convolution integral. These steps are shown below. In Fig. 2.8, part (a) shows the true Fourier transform of some function $x(t)$. In part (b), the transform is rotated about

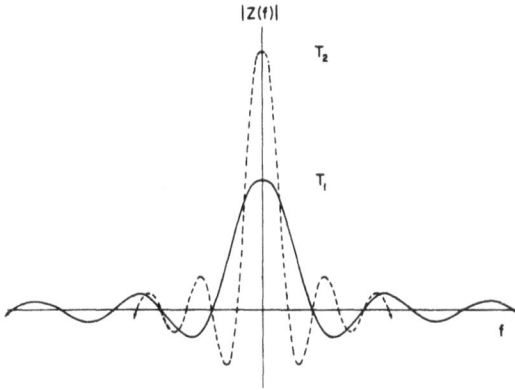

Fig. 2.7. Fourier transforms of boxcar functions
of durations T_1 and T_2.

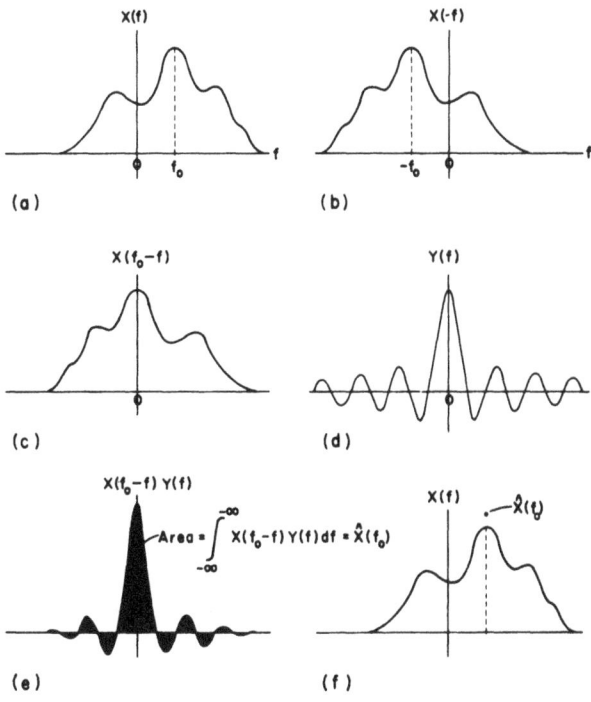

Fig. 2.8. Graphical evaluation of a convolution integral.

$f = 0$ to obtain the transform $X(-f)$. In part (c), the rotated transform is translated to the right by an amount f_0 so that the original transform value at f_0 now falls at $f = 0$. By performing this rotation and transformation, the first function $X(f_0 - f)$ in the above convolution has been obtained. Part (d) shows the Fourier transform $Y(f)$ of the boxcar function associated with the finite record length T. In part (e) of this figure, the product of the transforms shown in parts (c) and (d) is plotted. The area under this product curve is then equal to the value of the convolution integral evaluated at f_0. This is also equal to the finite transform at frequency f_0. In part (f), this value of the finite transform is plotted as a point at frequency f_0. The difference between this point and the true value of the transform at f_0 is the inaccuracy caused by taking the transform over the finite time interval T.

Values of the finite transform at frequency points other than f_0 are evaluated by substituting each new frequency point in place of f_0 in the convolution integral. For example,

$$\hat{X}(f_1) = \int_{-\infty}^{\infty} X(f_1 - f) Y(f) df. \tag{2.48}$$

This operation must be repeated for all the frequency points to be determined in the finite transform.

Fourier Series

Fourier analysis techniques are also used to describe periodic data that are not just simple sinusoids. Assume an arbitrary signal $x(t)$ that repeats itself exactly every T seconds:

$$x(t) = x(t + T) = x(t + 2T) \; . \; . \; . \; .$$

This time history can be described in terms of the infinite Fourier series,

$$x(t) = \frac{1}{T} \sum_{n=-\infty}^{\infty} \mathscr{X}\left(\frac{n}{T}\right) e^{j2\pi t \frac{n}{T}}, \qquad n = 1, 2, 3, \; . \; . \; . \tag{2.49}$$

where

$$\mathscr{X}\left(\frac{n}{T}\right) = \int_{-T/2}^{T/2} x(t) e^{-j2\pi t \frac{n}{T}} dt. \tag{2.50}$$

The resemblance of this discrete spectral quantity to the direct Fourier transform in Eq. (2.25) and to the finite direct Fourier transform of Eq. (2.44) is clear. Note the argument of the spectral quantity. These spectral values occur only at integer values of the frequency $1/T$. This frequency is called the fundamental, and n/T is its nth harmonic.

The relationship between the Fourier series in Eq. (2.49) and the inverse transform of Eq. (2.28) is not so simple. In fact, the formal proof of this relationship is complicated. Heuristically, the relationship can be demonstrated by taking the limits on $(1/T)$ and (n/T) as both T and n

approach infinity. As T becomes infinite, $1/T$ becomes an increment in a continuous variable;

$$\lim_{T \to \infty}\left(\frac{1}{T}\right) \to df.$$

Likewise, the limit of n/T must be taken so that

$$\lim_{\substack{n \to \infty \\ T \to \infty}}\left(\frac{n}{T}\right) \to f.$$

Then the summation will become the following integration:

$$\lim_{\substack{n \to \infty \\ T \to \infty}}\left[\frac{1}{T}\sum_{n=-\infty}^{\infty}\mathscr{X}\left(\frac{n}{T}\right)e^{-j2\pi t\frac{n}{T}}\right] \to \int_{-\infty}^{\infty}X(f)e^{-j2\pi tf}df. \qquad (2.51)$$

In many texts, the spectral function is defined slightly differently;

$$D_n = \frac{1}{T}\mathscr{X}\left(\frac{n}{T}\right) = \frac{1}{T}\int_{-T/2}^{T/2}x(t)e^{-j2\pi t\frac{n}{T}}\,dt, \qquad (2.52)$$

and in terms of the time function,

$$x(t) = \sum_{n=-\infty}^{\infty}D_n e^{j2\pi t\frac{n}{T}}. \qquad (2.53)$$

Notice that the only difference is that the division by the record length occurs in the computation of the spectral function instead of in the computation of the time function.

As with the Fourier integral transformation, there are a number of different forms of the Fourier series. The other two most common forms are derived below from the complex exponential series. The spectral function is

$$D_n = \frac{1}{T}\int_{-T/2}^{T/2}x(t)e^{-j2\pi t\frac{n}{T}}dt = \frac{1}{T}\int_{-T/2}^{T/2}x(t)\cos 2\pi t\frac{n}{T}\,dt$$

$$- j\frac{1}{T}\int_{-T/2}^{T/2}x(t)\sin 2\pi t\frac{n}{T}\,dt = a_n - jb_n. \qquad (2.54)$$

The time history, usually written in terms of positive frequency components, is

$$x(t) = a_0 + 2\sum_{n=1}^{\infty}\left(a_n \cos 2\pi t\frac{n}{T} + b_n \sin 2\pi t\frac{n}{T}\right)$$

$$a_0 = \frac{1}{T}\int_{-T/2}^{T/2}x(t)\,dt$$

$$a_n = \frac{1}{T}\int_{-T/2}^{T/2}x(t)\cos 2\pi t\frac{n}{T}\,dt$$

$$b_n = \frac{1}{T}\int_{-T/2}^{T/2}x(t)\sin 2\pi t\frac{n}{T}\,dt. \qquad (2.55)$$

This is the Fourier series in terms of the sines and cosines of the harmonics. The coefficients a_n and b_n are real, whereas D_n is a complex quantity.

By using the trigonometric identity

$$a \cos x + b \sin x = \sqrt{a^2 + b^2} \cos (x - \phi), \qquad (2.56)$$

the series can be obtained in terms of an amplitude value and a phase shift at each harmonic. Thus,

$$x(t) = a_0 + 2 \sum_{n=1}^{\infty} \left[c_n \cos \left(2\pi t \frac{n}{T} + \phi_n \right) \right], \qquad (2.57)$$

where

$$c_n = \sqrt{a_n^2 + b_n^2} \qquad (2.58a)$$

$$\phi_n = \tan^{-1} \left[\frac{b_n}{a_n} \right]. \qquad (2.58b)$$

In Fig. 2.9a, a discrete Fourier spectrum is plotted in terms of real and imaginary values. In part (b), it is plotted in terms of a modulus and phase.

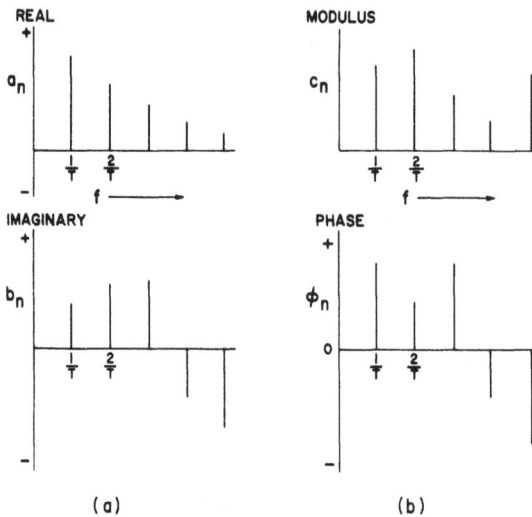

Fig. 2.9. Discrete Fourier transforms; (a) real and imaginary presentation, (b) modulus and phase presentation.

The primary value of the Fourier series is that it simplifies the calculations for the response of linear systems with complicated, but periodic, input functions. At each frequency contained in the input, the response is calculated by simple steady state (for sinusoidal inputs) techniques. This consists of simply scaling the magnitude and shifting the phase of the component of the input at that frequency. The response is calculated independently at each frequency. Then all the responses are summed vectorially to compute the total response.

2.4 Laplace Transforms

All of the preceding discussions have been devoted to Fourier transforms, as they are the most common ones used in the analysis of shock data. However, there are many occasions when Laplace transformation results in a simpler solution to certain problems. From Eq. (2.23), the equation of a linear integral transform is

$$T[x(t)] = \int_a^b K(t, u)x(t)dt. \tag{2.59}$$

If e^{-st} is used as the kernel function, the lower limit a is set to 0, and the upper limit b is set to infinity, the transform is known as the Laplace transform

$$\mathscr{L}[x(t)] = X(s) = \int_0^\infty e^{-st}x(t)dt, \tag{2.60}$$

where

$$\mathscr{L}[x(t)] = X(s) = \text{the Laplace transform.} \tag{2.61}$$

Notice that the differences between the Laplace and Fourier transforms are in (a) the lower limit and (b) the argument of the exponential kernel. The argument of the Laplace transformation kernel is a complex variable

$$s = \sigma + j2\pi f. \tag{2.62}$$

Thus this kernel is a damped version of the Fourier kernel (or, perhaps more properly, the Fourier kernel is the undamped portion of the Laplace kernel):

$$e^{-st} = e^{-\sigma t - j2\pi ft} = e^{-\sigma t} \cdot e^{-j2\pi ft}. \tag{2.63}$$

The inverse Laplace transform is

$$\mathscr{L}^{-1}[x(s)] = x(t) = \frac{1}{2\pi j} \int_{c-j\infty}^{c+j\infty} X(s)e^{ts}ds; \; \sigma_a < c, \tag{2.64}$$

where

$$\mathscr{L}^{-1}[x(s)] = \text{the inverse Laplace transform}$$

$$\sigma_a = \text{the minimum value of ``damping'' that will make the inverse transform converge.}$$

As with the Fourier transforms, a table of transform pairs can be used instead of performing the actual integrations. For tables, see Refs. 8 and 10.

The primary advantages of the Laplace transform over the Fourier transform are

● The Laplace integral converges for a large class of functions for which the Fourier integral is divergent, and

● Initial conditions can be introduced directly into the integral.

For example, consider the simple step function shown in Fig. 2.10. Strictly speaking, this function does not converge;

$$\int_{-\infty}^{\infty} u(t)dt = \int_{0}^{\infty} (1)dt. \qquad (2.65)$$

Therefore, its Fourier transform should not exist. The existence of the Fourier transform of this unit step function is usually "justified" by finding the transform of a decaying exponential that starts at time zero (see Fig. 2.11);

$$x(t) = \begin{cases} e^{-\alpha t} & 0 \leq t \\ 0 & 0 \geq t \end{cases} \qquad (2.66)$$

$$X(f) = \int_{0}^{\infty} e^{-\alpha t} e^{-j2\pi ft} dt$$

$$= \int_{0}^{\infty} e^{-(\alpha + j2\pi f)t} dt$$

$$= \frac{1}{\alpha + j2\pi f}. \qquad (2.67)$$

Then the limit is taken as $\alpha \to 0$ ($e^{-(0)t} = 1$) and the Fourier transform of the step function is defined as $1/(j2\pi f)$.

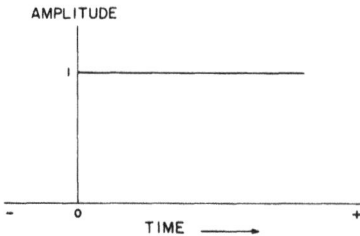

Fig. 2.10.   Unit step function.

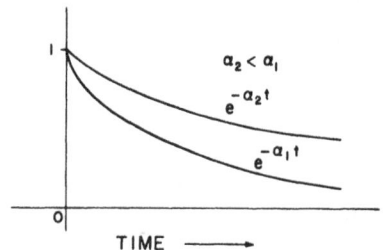

Fig. 2.11.   Decaying exponential functions.

This type of justification is not needed for the Laplace transform because the real part of the exponential kernel forces the transform to converge. The transform is found simply;

$$X(s) = \int_0^\infty u(t)e^{-st}dt$$

$$= \int_0^\infty e^{-st}dt$$

$$= \frac{1}{s}. \qquad (2.68)$$

By rewriting Eq. (2.68) as follows

$$X(s) = \int_0^\infty x(t)e^{-\sigma t}\cos 2\pi ft dt - j \int_0^\infty x(t)e^{-\sigma t}\sin 2\pi ft dt, \qquad (2.69)$$

it can more easily be seen that the Laplace transform kernel consists of damped trigonometric terms, whereas the Fourier transform kernel consists of undamped trigonometric terms. It is this damping that permits the convergence of Laplace transforms for some functions whose Fourier transforms diverge.

As previously stated, the second major advantage of the Laplace transform over the Fourier transform is the ease with which initial conditions are introduced. To show this, the Laplace transforms of the derivative and the integral of a function must be found;

$$\mathscr{L}\left[\frac{dx(t)}{dt}\right] = \int_0^\infty \frac{dx(t)}{dt}e^{-st}dt. \qquad (2.70)$$

The solution to the above integral can be found by integrating by parts,

$$\int u\,dv = uv - \int v\,du. \qquad (2.71)$$

Let $u = e^{-st}$ and $dv = [dx(t)/dt]dt$. Then $du = -se^{-st}dt$ and $v = x(t)$, and

$$\mathscr{L}\left[\frac{dx(t)}{dt}\right]\left[e^{-st}x(t)\right]_0^\infty + \int_0^\infty x(t)se^{-st}dt = -x(0) + s\int_0^\infty x(t)e^{-st}dt$$

$$= -x(0) + sX(s). \qquad (2.72)$$

Thus, the Laplace transform of a derivative of a function is equal to the Laplace transform of the function multiplied by the complex frequency value s, and from this product is subtracted the initial value of the function. (Care must be exercised to be sure that the initial value used, if it is not single valued, is the one obtained by approaching time zero from the right.) Transforms of higher order derivatives are found in the same manner;

$$\mathcal{L}\left[\frac{d^2x(t)}{dt^2}\right] = s^2X(s) - sx(0) - \frac{dx(0)}{dt}$$

and   (2.73)

$$\mathcal{L}\left[\frac{d^nx(t)}{dt^n}\right] = s^nX(s) - \sum_{i=1}^{n} s^{(n-i)}\left[\frac{d^{(i-1)}x(0)}{dt^{(i-1)}}\right].$$

The Laplace transform of the integral of a function can be found from the above properties of differentiation. Let

$$y(t) = \int x(t)dt, \quad Y(s) = \mathcal{L}\left[\int x(t)dt\right],$$

and   (2.74)

$$x(t) = \frac{dy(t)}{dt} = \frac{d\left[\int x(t)dt\right]}{dt}.$$

Therefore the transform of $x(t)$ is

$$X(s) = sY(s) - y(0).$$ (2.75)

The transform of the integral is found simply by rearranging;

$$Y(s) = \frac{X(s)}{s} - \frac{y(0)}{s}$$

$$\mathcal{L}\left[\int x(t)dt\right] = \frac{X(s)}{s} + \frac{1}{s}\int_{-\infty}^{0} x(t)dt.$$ (2.76)

In words, the Laplace transform of the integral of a function is equal to the Laplace transform of the function plus the value of the integral at time zero, the quantity divided by the complex frequency variable s. (Again, the integral must be evaluated by approaching time zero from the right.) Higher order integrals are found by the following formula:

$$\mathcal{L}\left[\int n\int x(t)dt^n\right] = \frac{X(s)}{s^n} + \sum_{i=1}^{n}\left[\frac{\int i\int x(0)dt^i}{s^{(n-i+1)}}\right].$$ (2.77)

Since the transforms of derivatives and integrals of functions involve initial conditions, the mechanism is available for entering these initial conditions directly into the transforms of differential or integro-differential equations.*

As an example, consider the differential equation

$$\frac{d^2x(t)}{dt^2} + x(t) = 0$$ (2.78)

with initial conditions of $[dx(0)]/dt = 0$ and $x(0) = 1$.

---

*Integrodifferential equations contain both integrals and derivatives.

Transforms are taken of the differential equation

$$\mathscr{L}\left[\frac{d^2x(t)}{dt^2}\right] + \mathscr{L}[x(t)] = \mathscr{L}[0]$$

$$\left[s^2X(s) - sx(0) - \frac{dx(0)}{dt}\right] + [X(s)] = 0$$

$$s^2X(s) + X(s) \underbrace{- s - 0}_{\text{initial conditions}} = 0$$

$$X(s) = \frac{s}{s^2 + 1}. \tag{2.79}$$

From a table of transform pairs, the inverse transform is found:

$$x(t) = \mathscr{L}^{-1}[X(s)] = \mathscr{L}^{-1}\left[\frac{s}{s^2 + 1}\right] = \cos t. \tag{2.80}$$

Chapter 3

RESPONSE OF LINEAR SYSTEMS

Since the goal of any transient measurement is to examine the response of some system, this chapter will examine various methods for determining the response of systems. The systems discussed are restricted to ideal linear systems. Methods for calculating the response in the time domain, the frequency domain, and the complex frequency domain are described. In addition, methods for computing the overall response of systems connected in tandem are explained.

3.1 Definitions

The analysis procedures to be described in the following sections of this chapter assume that the physical system under consideration is linear, that its fundamental properties do not change with time, and that there is a single input to and a single output from the system. Such a system is usually termed an ideal system, or in more specific terms, it is called a single input, single output, constant-parameter, linear system.

The term linearity denotes two basic characteristics of the response of the system:

● The response is additive, and
● The response is homogeneous.

The first of these characteristics implies that the response of the system to the sum of several excitations is equal to the sum of the system responses obtained when each excitation is applied individually. This is shown in Fig. 3.1a. The excitations, or inputs, are designated by x's, and the responses, or outputs, are designated by y's.

Homogeneity means that the response of the system to the product of some constant and the excitation is equal to the product of this same constant and the response generated by the excitation alone. The homogeneity requirement is depicted in Fig. 3.1b and the total linearity requirement is shown in Fig. 3.1c.

The constant-parameter requirement means that the integrodifferential equations relating the output of the system to its input must be of the constant-parameter type. The single input and output require-

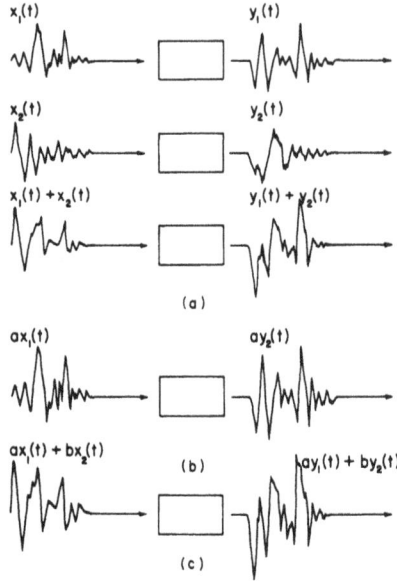

Fig. 3.1.   Linearity definition; (a) additive
rule, (b) homogeneity, (c) linearity.

ments simply require that there must be only one input to and only one
output from the system.

With actual physical systems, these restrictions are never met exactly.
There are no truly ideal systems. In fact, no practical system can truly
meet any of the requirements of the ideal system. There are no systems
that are linear over the entire possible range of input levels. Neither
are there any physical systems that are truly invariant with time. In
fact, there are no physical systems with only a single input and only a
single output.

However, the single input, single output, constant-parameter, linear
system assumption is valid for a vast number of physical systems. The
key element is that this model for the system must allow the calculation
of the response within the required degree of accuracy. In other words,
the response need not be exactly perfect – only accurate to the degree
that matches its engineering application.

It is important to keep in mind that, for the physical system to be
modeled by the ideal system, the deviations from the stated requirements must be minor, not intentional. For example, an amplifier that
deviates by ± 1 percent from linearity over its rated output range would
meet the linearity requirement, but a squaring circuit would definitely
violate the linearity requirement. Slow, temperature-related drifts
in the damping of a transducer would not preclude the use of an ideal
system model for short-duration measurements. However, a frequency

modulation circuit, where the reactance of a tuned circuit is varied in proportion to the input level, definitely violates the constant-parameter requirement. Noise on the input of an amplifier can be neglected if the signal level is much higher than the noise level. However, the vibration response of a circuit board, connected mechanically to the structure at four points and having an acoustic coupling, definitely violates the single input requirement. Heat dissipated by an amplifier is usually a minor deviation from the single output requirement, but a fluid reservoir having several output pipes would be a major deviation.

3.2   Response Calculations in the Time Domain

Assuming that the system does qualify as a simple linear system, the response can be calculated in either the time or frequency domain. The time domain calculation will be discussed first. The output is expressed as a function of the input and the characteristics of the system by a convolution equation;

$$y(t) = \int_{-\infty}^{\infty} x(\tau)h(t-\tau)d\tau \quad \text{or} \quad \int_{-\infty}^{\infty} x(t-\tau)h(\tau)d\tau, \qquad \text{(3.1a), (3.1b)}$$

where

$y(t)$ = the output

$x(t)$ = the input

$h(t)$ = the weighting, or unit impulse response, function.

The *weighting function* is the time response (output) of a system to a delta function input (an infinite amplitude, zero duration, unit area pulse). Figure 3.2 illustrates the measurement of the weighting function. The output of the system is by definition the weighting function when the input is a delta function.

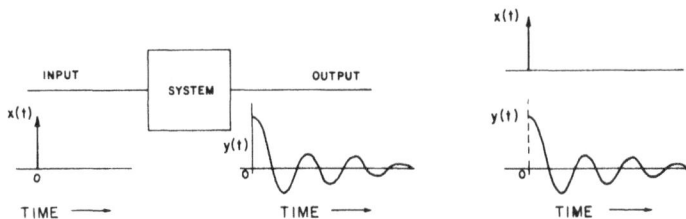

Fig. 3.2.   Illustration of the weighting function.

In practice, there are three methods used to measure the weighting function directly:

1. A high-level pulse, whose duration is short compared to the period of the upper cutoff frequency of the system, is applied.

2. A step function input is applied and the output of the system is differentiated.

3. A random noise input whose spectral density is flat is applied, and the crosscorrelation function between the input and output is computed. The input spectrum should include the system passband, and in fact contain frequencies much higher than the upper cutoff frequency of the system.

The first two methods frequently suffer from signal-to-noise problems. The amplitude of the input pulse must be restricted if the system is to operate in its linear range; hence, the output is frequently very small. By applying a step function, a larger output level can be obtained from the system; however, noise in the differentiation may result in very little if any signal-to-noise ratio improvement after the output is differentiated. The last method is usually the most practical. The manner in which the crosscorrelation function yields the weighting function will be demonstrated. The crosscorrelation function is defined as follows:

$$R_{xy}(\tau) = \lim_{T \to \infty} \frac{1}{T} \int_{-T/2}^{T/2} x(t)y(t+\tau)dt, \qquad (3.2)$$

where

$\quad R_{xy}(\tau) =$ the crosscorrelation function at a delay value of τ

$\quad x(t) =$ the input to the system

$\quad y(t) =$ the output of the system.

The output can be expressed in terms of the convolution of the input and the weighting function. From Eq. (3.1),

$$y(t+\tau) = \int_{-\infty}^{\infty} h(u)x(t+\tau-u)du.$$

Now

$$R_{xy}(\tau) = \lim_{T \to \infty} \frac{1}{T} \int_{-T/2}^{T/2} x(t) \left[\int_{-\infty}^{\infty} h(u)x(t+\tau-u) \right] dt. \qquad (3.3)$$

Changing the order of integration yields

$$R_{xy}(\tau) = \int_{-\infty}^{\infty} h(u) \left[\lim_{T \to \infty} \frac{1}{T} \int_{-T/2}^{T/2} x(t)x(t+\tau-u)dt \right] du. \qquad (3.4)$$

Note that

$$\lim_{T \to \infty} \frac{1}{T} \int_{-T/2}^{t/2} x(t)x(t+\tau-u)dt = R_{xx}(\tau-u), \qquad (3.5)$$

where

$R_{xx}(\tau-u) =$ the autocorrelation function of the input at a delay value of $(\tau-u)$

$$R_{xy}(\tau) = \int_{-\infty}^{\infty} h(u)R_{xx}(\tau - u)du. \tag{3.6}$$

If the input signal is white noise, its autocorrelation function is a delta function

$$R_{xx}(\tau) = \delta(\tau).$$

Hence,

$$R_{xy}(\tau) = \int_{-\infty}^{\infty} h(u)\delta(\tau - u)du, \tag{3.7}$$

where $\delta(\tau - u)$ is the delta function at $u = \tau$. Then

$$R_{xy}(\tau) = h(\tau). \tag{3.8}$$

The weighting function also describes two important properties of the system. First, the weighting function can be used to determine whether the system is stable. For the system to be stable, the integral of the weighting function over infinite limits must be finite;

$$\int_{-\infty}^{\infty} h(\tau)d\tau < \infty. \tag{3.9}$$

For the system to be physically realizable, it must respond only to previously applied inputs. It cannot anticipate inputs to be applied in the future. This means that the weighting function must be zero when its argument has negative values. In equation form,

$$h(t) = 0, \text{ for } t < 0. \tag{3.10}$$

The weighting function can be thought of as a window through which the input is viewed. If, for example, the weighting function is a simple boxcar function of duration T, the output is the integral of the input from the time the output is desired, back T seconds. Each value in this interval is weighted equally, and values of the input applied more than T seconds previously do not influence the output. If the weighting function is an exponential, values of the input that have just occurred contribute more heavily to the output than values that have previously occurred. Theoretically, all past inputs contribute to the output, but for practical purposes, there is some value of time for which the weighting function has decayed to such a small value that inputs applied prior to that time contribute negligibly to the output.

A graphical method of performing the convolution was illustrated in Chapter 2 in the section titled, "Finite Fourier Transforms." Generally, this integral is difficult to evaluate.

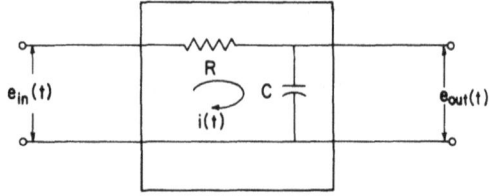

Fig. 3.3.   An RC low-pass filter.

As an example of the calculation of the weighting function of a simple linear system, consider the RC low-pass filter in Fig. 3.3. The sum of the voltage rises equals the sum of the voltage drops;

$$e_{in}(t) = i(t)R + \frac{1}{C}\int i(t)dt. \tag{3.11}$$

For convenience, convert from current to charge; $q(t) = \int i(t)dt$. Then

$$e_{in}(t) = q(t)R + \frac{q(t)}{C}$$

and   (3.12)

$$e_{out}(t) = \frac{q(t)}{C}.$$

Assume that $e_{in}(t)$ is a unit step function where $e_{in}(t) = 1$. Then

$$dq = \frac{1}{R}(1 - q/C)dt$$

(3.13)

$$\frac{dq}{1 - q/C} = \frac{1}{R}dt.$$

Let $u = 1 - q/C$ and $du = -(1/C)dq$. Eq. (3.13) becomes

$$\frac{-Cdu}{u} = \frac{1}{R}dt$$

$$-\ln u = \frac{1}{RC}t + k$$

$$-\ln(1 - q/C) = \left(\frac{1}{RC}\right)t + k$$

$$1 - q/C = e^{-\left(\frac{t}{RC} + k\right)}$$

$$q = C\left[1 - e^{-\left(\frac{t}{RC} + k\right)}\right] \tag{3.14}$$

$$e_{out}(t) = q/C = 1 - e^{-\left(\frac{t}{RC} + k\right)}$$

Since $e_{out}(t)$ at $t=0$ is zero, $k=0$. Thus,

$$e_{out}(t) = 1 - e^{-\frac{t}{RC}}. \qquad (3.15)$$

To obtain the unit impulse response $h(t)$, the response to the step function must be differentiated. This yields the result

$$h(t) = \frac{d[e_{out}(t)]}{dt} = -\left(e^{-\frac{t}{RC}}\right)\left(-\frac{1}{RC}\right) = \left(\frac{1}{RC}\right)e^{-\frac{t}{RC}}. \qquad (3.16)$$

3.3 Response Calculations in the Frequency Domain

By restricting the linear system to being stable and physically realizable, it is possible to relate the output of the system to its input in the frequency domain by means of the frequency response function

$$Y(f) = X(f)H(f), \qquad (3.17)$$

where

$Y(f)$ = the Fourier transform of the output

$X(f)$ = the Fourier transform of the input

$H(f)$ = the frequency response function. Note that it is also the Fourier transform of the weighting function.

By working in the frequency domain, the difficult convolution operation is replaced by a simple multiplication operation. Also of significant importance is the fact that what happens at one frequency is independent of what happens at any other frequency. For example, suppose the response has been calculated for a certain input, and the magnitude of the response at some frequency f_0 is the only response parameter of interest. If a new signal is added to the input, the level of the response at f_0 will not change as long as the new signal does not contain energy at frequency f_0.

The frequency response function is a complex quantity. Figure 3.4 ·shows two ways in which it can be displayed. Part (a) shows the real and imaginary components and part (b) shows the modulus, called gain factor, and the phase factor. The latter presentation is more common.

There are three methods commonly employed to measure the frequency response function of a system. In the first method, a sine wave is applied to the input of the system. Because of the restrictions placed on the system, the output must be a sine wave at the same frequency. Thus the frequency response function at that frequency is simply the ratio of the output to the input (gain factor) and the phase shift between the output and the input (phase factor). By varying the frequency of the sine wave and repeating the ratio and phase shift measurements, the entire frequency response function can be measured.

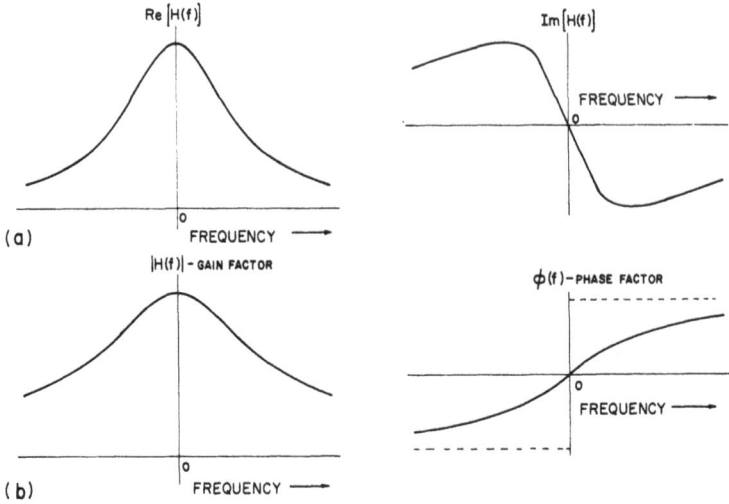

Fig. 3.4.   Frequency response function.

The second method consists of applying a transient to the system and measuring the Fourier transforms of both the input transient and the response of the system;

$$H(f) = \frac{Y(f)}{X(f)}. \tag{3.18}$$

The third method consists of applying a random noise signal with a flat spectral density over the frequency range of the system. Then the cross-spectral density between the input and output of the system is computed. The cross-spectral density function is defined by

$$G_{xy}(f) = \lim_{T \to \infty} \frac{1}{T} X^*(f)Y(f), \tag{3.19}$$

where

$$G_{xy}(f) = \text{the cross spectral density function}$$
$$X^*(f) = \text{the complex conjugate of } X(f).$$

Combining Eqs. (3.17) and (3.19) yields

$$G_{xy}(f) = \lim_{T \to \infty} \frac{1}{T} X^*(f)X(f)H(f). \tag{3.20}$$

Since the definition of the ordinary spectral density function is

$$G_{xx}(f) = \lim_{T \to \infty} \frac{1}{T} X^*(f)X(f), \tag{3.21}$$

where $G_{xx}(f)$ = the ordinary spectral density function, Eq. (3.20) becomes

$$G_{xy}(f) = G_{xx}(f)H(f).$$ (3.22)

Since the spectral density function of the input is a constant over the passband of the system,

$$H(f) = \frac{G_{xy}(f)}{k},$$ (3.23)

where

k = the spectral density of the input.

Of the three methods, the first is the easiest and most accurate to implement, although it may be the most time-consuming.

To illustrate the frequency response function application, the frequency response function of the simple mechanical oscillator in Fig. 3.5 will be found, and then its response to a particular time history will be calculated.

Fig. 3.5. Simple mechanical oscillator.

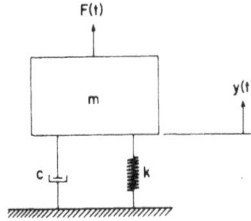

From Nexton's second law it is known that the sum of the forces applied to the mass, including its inertial force, must be zero. Thus the equation of motion can be written

$$F(t) - m\frac{d^2y(t)}{dt^2} - c\frac{dy(t)}{dt} - ky(t) = 0,$$ (3.24)

where

$F(t)$ = the force applied to the mass

m = the mass

k = the spring constant

c = the damping constant

$y(t)$ = the displacement of the mass (the base is fixed).

Let the frequency response function relating the displacement of the mass to the applied force be calculated. Since the solution is purely analytical, a simplification can be employed. If the input force is made a delta function, the displacement output will be the weighting function.

The Fourier transform of the displacement output will be the desired frequency response function.

The Fourier transform of Eq. (3.24) when $F(t)$ is a delta function is

$$1 - m[-(2\pi f)^2 H(f)] - c[j2\pi f H(f)] - k[H(f)] = 0. \qquad (3.25)$$

Since

$$F[\delta(t)] \quad = 1$$

$$F[y(t)] \quad = H(f)$$

$$F\left[\frac{dy(t)}{dt}\right] = j2\pi f H(f)$$

$$F\left[\frac{d^2 y(t)}{dt^2}\right] = -(2\pi f)^2 H(f). \qquad (3.26)$$

Solving Eq. (3.25) for $H(f)$ gives

$$H(f) = \frac{1}{k - (2\pi f)^2 m + j2\pi f c}.$$

The gain and phase factors are frequently normalized as follows:

$$|H(f)| = \frac{(1/k)}{\sqrt{[1 - (f/f_n)^2]^2 + [2\zeta(f/f_n)]^2}}$$

$$\phi(f) = \tan^{-1}\left[\frac{2\zeta(f/f_n)}{1 - (f/f_n)^2}\right], \qquad (3.27)$$

where

$$\zeta = \frac{c}{2\sqrt{km}} = \text{the critical damping ratio}$$

$$f_n = \frac{1}{2\pi}\sqrt{k/m} = \text{the undamped natural frequency.}$$

To use the frequency response function, the Fourier transform is taken of the input;

$$F[x(t)] = X(f). \qquad (3.28)$$

This is multiplied by the frequency response function to yield the Fourier transform of the output;

$$Y(f) = X(f)H(f). \qquad (3.29)$$

If the time history of the output is desired, then the inverse Fourier transform is taken of the output;

$$F^{-1}[Y(f)] = y(t). \qquad (3.30)$$

3.4 Response Calculations in the Complex Frequency Domain

The Laplace transform quantity similar to the frequency response function is the *transfer function*. It is the Laplace transform of the weighting function

$$H(s) = \int_0^\infty h(t)e^{-st}dt, \qquad (3.31)$$

where

$$H(s) = \text{the transfer function.}$$

Its use is quite analogous to the use of the frequency response function. The response of a system in the complex frequency domain can be found simply by the product of the Laplace transform of the input and the transfer function,

$$Y(s) = X(s)H(s). \qquad (3.32)$$

The response time history can then be calculated by taking the inverse Laplace transform of the response,

$$y(t) = \mathcal{L}^{-1}[Y(s)]. \qquad (3.33)$$

In fact, if the primary goal is to determine the response time history of a system rather than its spectral composition, the use of Laplace transforms is preferable to the use of Fourier transforms.

The operations that must be performed to calculate the response time history are quite straightforward. The most difficult part comes in arranging the output in a form that simplifies the inverse transformation. Because the output will normally consist of the ratio of polynomials

$$Y(s) = \frac{N(s)}{D(s)}, \qquad (3.34)$$

where

$$N(s) = \text{the polynomial in the numerator of } Y(s)$$
$$D(s) = \text{the polynomial in the denominator of } Y(s),$$

this step consists of dividing the numerator by the denominator in order to express the output in terms of the quotient and a remainder;

$$Y(s) = N_1(s) + \frac{N_2(s)}{D(s)}, \qquad (3.35)$$

where

$$\frac{N_2(s)}{D(s)} = \text{a proper fraction.}$$

Next, the remainder is expanded in a partial fraction series. To factor the denominator, its roots must be found. (Roots are the values of s that satisfy the equation $D(s) = 0$.) In general, this is the most difficult operation involved in the solution. For an explanation of various root solving methods, see Ref. 12. The roots of the denominator are called poles. They are the complex frequency values that cause the transform of the output to be infinite. (The values of frequency that satisfy the equation $N(s)=0$ are known as zeros. They cause the value of the transform of the output to be zero at these frequencies.)

Assume that the output is expressed in the following manner:

$$Y(s) = \frac{n_k s^k + n_{k-1} s^{k-1} + \ldots + n_1 s + n_0}{s^j + d_{j-1} s^{j-1} + \ldots + d_1 s + d_0}. \tag{3.36}$$

First, the division is performed;

$$Y(s) = a_{k-j} s^{k-j} + a_{k-j-1} s^{k-j-1} + \ldots + a_1 s + a_0 + \frac{N_2(s)}{D(s)}. \tag{3.37}$$

Next, the roots of the denominator are found. This permits the denominator to be factored;

$$D(s) = (s - r_1)(s - r_2) \ldots (s - r_j), \tag{3.38}$$

where $r_j =$ the jth root.

Then the remainder is written as a series of partial fractions,

$$\frac{N_2(s)}{D(s)} = \frac{b_1}{(s - r_1)} + \frac{b_2}{(s - r_2)} + \ldots + \frac{b_j}{(s - r_j)}. \tag{3.39}$$

The next step is to find the values of the coefficients b_1, b_2, etc. To do this, the partial fraction expansion must be separated into two parts, one containing nonrepeated roots and the other containing repeated roots.

$$\frac{N_2(s)}{D(s)} = \left\{ \frac{b_1}{(s - r_1)} + \frac{b_2}{(s - r_2)} + \ldots + \frac{b_y}{(s - r_y)} \right\}$$

nonrepeated roots

$$+ \left\{ \frac{b_h}{(s - r_h)^u} + \ldots + \frac{b_j}{(s - r_j)^v} \right\}. \tag{3.40}$$

repeated roots

The techniques for finding the coefficients differ for repeated and nonrepeated roots. First consider any nonrepeated root r_e. If the entire proper fraction is multiplied by the factor $(s - r_e)$, then the value of b_e can be found by taking the limit as s approaches r_e;

$$\lim_{s \to r_e} \left\{ (s - r_e) \left(\frac{N_2(s)}{D(s)} \right) \right\} = \lim_{s \to r_e} \left[\frac{(s - r_e) b_1}{(s - r_1)} + \frac{(s - r_e) b_2}{(s - r_2)} + \ldots b_e + \ldots \right.$$

$$\left. + \frac{(s - r_e) b_y}{(s - r_y)} + \frac{(s - r_e) b_n}{(s - r_n)^u} + \ldots + \frac{(s - r_e) b_j}{(s - r_j)^v} \right] = b_e. \tag{3.41}$$

This technique essentially consists of dropping the factor containing the root under investigation from the denominator and then evaluating the remaining terms at the frequency equal to the root.

For example,

$$\frac{N_2(s)}{D(s)} = \frac{(s+1)}{s(s-3)(s+2)} = \frac{b_1}{s} + \frac{b_2}{(s-3)} + \frac{b_3}{(s+2)}$$

$$b_1 = \lim_{s \to 0} \left[\frac{sN_2(s)}{D(s)} \right] = \lim_{s \to 0} \left[\frac{(s+1)}{(s-3)(s+2)} \right] = \frac{1}{(-3)(2)} = -\frac{1}{6}$$

$$b_2 = \lim_{s \to 3} \left[\frac{(s-3)N_2(s)}{D(s)} \right] = \lim_{s \to 3} \left[\frac{(s+1)}{s(s+2)} \right] = \frac{(3+1)}{3(3+2)} = \frac{4}{15}$$

$$b_3 = \lim_{s \to -2} \left[\frac{(s+2)N_2(s)}{D(s)} \right] = \lim_{s \to -2} \left[\frac{(s+1)}{s(s-3)} \right] = \frac{-2+1}{-2(-2-3)} = \frac{-1}{10}$$

$$\frac{N_2(s)}{D(s)} = -\frac{1/6}{s} + \frac{4/15}{(s-3)} - \frac{1/10}{(s+2)}. \tag{3.42}$$

Another method of evaluating the coefficients of nonrepeated linear factors can be derived from Eq. (3.41);

$$\lim_{s \to r_e} \left[\frac{(s-r_e)N_2(s)}{D(s)} \right] = \lim_{s \to r_e} \left[\frac{(s-r_e)}{D(s)} \right] \lim_{s \to r_e} [N_2(f)]. \tag{3.43}$$

The first term in the above product is indeterminate (0/0). To evaluate this term, l'Hospital's rule is used;

$$\lim_{s \to r_e} \left[\frac{(s-r_e)}{D(s)} \right] = \lim_{s \to r_e} \left[\frac{[d(s-r_e)]/ds}{[dD(s)]/ds} \right] = \lim_{s \to r_e} \left[\frac{1}{\frac{d}{ds}[D(s)]} \right] = \frac{1}{\frac{d}{ds}[D(r_e)]}. \tag{3.44}$$

And since $\lim_{s \to r_e} [N_2(s)] = N_2(r_e)$,

$$b_e = \lim_{s \to r_e} \left[\frac{(s-r_e)N_2(s)}{D(s)} \right] = \frac{N_2(r_e)}{\frac{d}{ds}[D(r_e)]}. \tag{3.45}$$

In words, the coefficient of a nonrepeated factor is the quotient of the numerator of the entire proper fraction evaluated at the root associated with the factor divided by the derivative of the denominator of the entire proper fraction, evaluated at the same root.

Consider the previous example:

$$\frac{N_2(s)}{D(s)} = \frac{(s+1)}{s(s-3)(s+2)} = \frac{b_1}{s} + \frac{b_2}{(s-3)} + \frac{b_3}{(s+2)}$$

$$\frac{d}{ds}[D(s)] = (s-3)(s+2) + s(s-3) + s(s+2)$$

$$b_1 = \frac{N_2(0)}{\frac{d}{ds}[D(0)]} = \frac{1}{(-3)(+2)+0+0} = -\frac{1}{6}$$

$$b_2 = \frac{N_2(3)}{\frac{d}{ds}[D(3)]} = \frac{3+1}{(3-3)(3+2)+(3)(3-3)+3(3+2)} = \frac{4}{15}$$

$$b_3 = \frac{N_2(-2)}{\frac{d}{ds}[D(-2)]}$$

$$= \frac{-2+1}{(-2-3)(-2+2)+(-2)(-2-3)+(-2)(-2+2)} = \frac{-1}{10}.$$

$$(3.46)$$

To evaluate repeated factors, a different method must be used. Consider the term $b_j/[(s-r_j)^v]$ in Eq. (3.47). This term must be further expanded in a partial fraction series,

$$\frac{b_j}{(s-r_j)^v} = \frac{c_1}{(s-r_j)} + \frac{c_2}{(s-r_j)^2} + \ldots + \frac{c_v}{(s-r_j)^v}. \qquad (3.47)$$

To find the value of the coefficient c_1, c_2, etc., associated with a repeated root, the entire proper fraction must be multiplied by the repeated factor $(s-r_j)^v$;

$$(s-r_j)^v \frac{N_2(s)}{D(s)} = c_1(s-r_j)^{v-1} + c_2(s-r_j)^{v-2} + \ldots + c_v + (s-r_j)^v \left[\frac{b_1}{(s-r_j)} \right.$$

$$\left. + \frac{b_2}{(s-r_2)} + \ldots + \frac{b_g}{(s-r_g)} + \frac{b_n}{(s-r_n)^u} + \frac{b_{v-1}}{(s-r_{v-1})^m} \right]. \qquad (3.48)$$

By letting s approach r_j,

$$\lim_{s \to r_j} \left[\frac{(s-r_j)^v N_2(s)}{D(s)} \right] = c_v. \qquad (3.49)$$

To find the $(v-i)$th coefficient, the bracketed quantity in the above equation is differentiated i times with respect to frequency. Then this quantity is divided by i factorial and is evaluated at a frequency of r_j;

$$c_{v-1} = \lim_{s \to r_j} \frac{d}{ds} \left[\frac{(s-r_j)^v N_2(s)}{D(s)} \right] \left[\frac{1}{i!} \right]$$

$$c_{v-2} = \lim_{s \to r_j} \frac{d^2}{ds^2} \left[\frac{(s-r_j)^v N_2(s)}{D(s)} \right] \left(\frac{1}{2!} \right)$$

. .
. .
. .

$$c_1 = \lim_{s \to r_j} \frac{d^{v-1}}{ds^{v-1}} \left[\frac{(s-r_j)^v N_2(s)}{D(s)} \right] \left[\frac{1}{(v-1)!} \right]. \tag{3.50}$$

The computation of these coefficients of the partial fractions expansion is, in fact, the computation of the residues at each pole as stated in complex variable theory, Ref. 13.

As an example of these computations, consider the equation

$$\frac{N_2(s)}{D(s)} = \frac{1}{(s+3)(s-2)^3} = \frac{b_1}{(s+3)} + \frac{c_1}{(s-2)} + \frac{c_2}{(s-2)^2} + \frac{c_3}{(s-2)^3}. \tag{3.51}$$

From Eq. (3.51),

$$b_1 = \lim_{s \to -3} \left\{ (s+3) \left[\frac{1}{(s+3)(s-2)^3} \right] \right\} = \lim_{s \to -3} \left[\frac{1}{(s-2)^3} \right] = \left[\frac{1}{(-3-2)^3} \right] = \frac{-1}{125}. \tag{3.52}$$

From Eq. (3.51),

$$c_3 = \lim_{s \to 2} \left\{ (s-2)^3 \left[\frac{1}{(s+3)(s-2)^3} \right] \right\} = \lim_{s \to 2} \left[\frac{1}{(s+3)} \right] = \left[\frac{1}{2+3} \right] = \frac{+1}{5}. \tag{3.53}$$

From Eq. (3.51),

$$c_2 = \lim_{s \to 2} \frac{d}{ds} \left[\frac{1}{(s+3)} \right] = \lim_{s \to 2} \left[\frac{-1}{(s+3)^2} \right] = \left[\frac{-1}{(2+3)^2} \right] = \frac{-1}{25}.$$

$$c_1 = \lim_{s \to 2} \frac{d^2}{ds^2} \left[\frac{1}{(s+3)} \right] = \lim_{s \to 2} \left[\frac{2}{(s+3)^3} \right] = \left[\frac{2}{(2+3)^3} \right] = \frac{2}{125}. \tag{3.54}$$

Therefore,

$$\frac{N_2(s)}{D(s)} = \frac{1/125}{(s+3)} + \frac{1/5}{(s-2)} - \frac{1/25}{(s-2)^2} + \frac{2/125}{(s-2)^3}. \tag{3.55}$$

The examples have employed real values for the roots, but complex values work equally well.

Now that the equation representing the system output has been expanded, the next and final step consists of taking the inverse transform.

Because a linear transformation is being used, the total inverse transform is equal to the sum of the individual inverse transforms.

The inverse transformation of terms containing nonnegative powers of s results in impulses for time functions (see Ref. 8, p. 256). The inverse transform of a constant is that constant times a delta function at time $t=0$;

$$\mathcal{L}^{-1}[k] = k\delta(t). \tag{3.56}$$

The inverse transform of s is a unit doublet impulse at time zero. Higher order powers of s yield higher order impulses. For more information, see Ref. 8.

The proper fraction terms are much more common. The inverse transform of a nonrepeated factor is

$$\mathcal{L}^{-1}\left[\frac{b_1}{(s-r_1)}\right] = b_1 e^{r_1 t}, \tag{3.57}$$

and the inverse transform of a repeated factor is

$$\mathcal{L}^{-1}\left[\frac{b_j}{(s-r_j)^v}\right] = \mathcal{L}^{-1}\left[\frac{c_1}{(s-r_j)^2} + \frac{c_2}{(s-r_j)^2} + \cdots + \frac{c_v}{(s-r_j)^v}\right]$$

$$= e^{r_j t}\left[c_1 + c_2 t + \frac{c_3 t^2}{(3-1)!} + \cdots + \frac{c_v t^{v-1}}{(v-1)!}\right]$$

$$= e^{r_j t}\sum_{i=1}^{v}\frac{c_i t^{i-1}}{(i-1)!} \quad \text{(note that } 0! = 1). \tag{3.58}$$

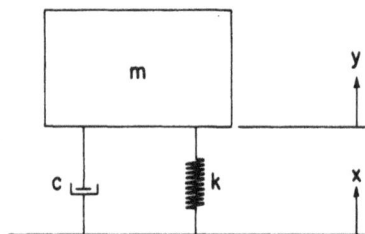

Fig. 3.6. Seismic system of an accelerometer.

An example will be worked to illustrate the use of the transform technique to determine a response time history. The seismic system of a typical accelerometer is shown in Fig. 3.6. The accelerometer provides a voltage output that is proportional to force in the spring element. This force is in turn proportional to the difference in displacement of the mass and the base of the transducer. The transfer function between the acceleration of the base and the displacement of mass relative to the base can be found by first writing the equation of motion;

$$-m\frac{d^2y(t)}{dt^2} - c\left[\frac{dy(t)}{dt} - \frac{dx(t)}{dt}\right] - k[y(t) - x(t)] = 0, \tag{3.59}$$

where

$x(t)$ = the displacement of the base of the accelerometer

$y(t)$ = the displacement of the mass of the accelerometer.

Separating the variables,

$$m\frac{d^2y(t)}{dt^2} + c\frac{dy(t)}{dt} + ky(t) = c\frac{dx(t)}{dt} + kx(t). \tag{3.60}$$

Then the Laplace transforms are taken of both sides of the equation,

$$m[s^2Y(s) - sy(0) - \dot{y}(0)] + c[sY(s) - y(0)] + kY(0) = c[sX(s) - x(0)] + kX(s)$$

$$ms^2Y(s) + csY(s) + kY(s) = csX(s) + kX(s) - cX(0) + msy(0) + m\dot{y}(0) + cy(0)$$

$$Y(s) = \frac{(cs+k)[X(s)]}{ms^2+cs+k} + \frac{m[y(0)+\dot{y}(0)] + c[y(0)-x(0)]}{ms^2+cs+k} \tag{3.61}$$

$$Y(s) = \left[\frac{cs+k}{ms^2+cs+k}\right]X(s), \text{ if the initial conditions are all zero.}$$

The transfer function desired is the one for a relative displacement output and a base acceleration input. This is

$$H(s) = \frac{\mathcal{L}[y(t) - x(t)]}{\mathcal{L}\left[\frac{d^2[x(t)]}{dt^2}\right]} = \frac{Y(s) - X(s)}{s^2X(s)}, \tag{3.62}$$

again assuming that the seismic system in the accelerometer is initially at rest.

Substituting Eq. (3.62) into Eq. (3.61), the transfer function of the accelerometer is found as follows:

$$H(s) = \frac{\left(\frac{cs+k}{ms^2+cs+k}\right)[X(s)] - X(s)}{s^2X(s)}$$

$$H(s) = \frac{-1}{s^2 + \frac{c}{m}s + \frac{k}{m}}, \tag{3.63}$$

or, as commonly normalized,

$$H(s) = \frac{-1}{s^2 + 2\zeta\omega_n s + \omega_n^2}, \tag{3.64}$$

where

$\zeta = \dfrac{c}{2\sqrt{km}}$ = the fraction of critical damping of the seismic system in the accelerometer

$\omega_n = \sqrt{k/m}$ =the undamped natural frequency in radians per second of the seismic system in the accelerometer.

To depart for a moment from the original goal of this example, the frequency response function of the accelerometer will be examined. Since in practice the frequency response function of the accelerometer would be measured with undamped sinusoids, it is legitimate to examine the transfer function along the line $s=j2\pi f$. Substituting $j2\pi f$ for s in Eq. (3.64) yields

$$H(2\pi f) = \frac{-1}{-(2\pi f)^2 + 2\zeta\omega_n(2\pi f) + \omega_n^2}. \qquad (3.65)$$

The magnitude (gain factor) and phase shift (phase factor) of this function are plotted in Fig. 3.7 for various fractions of critical damping.

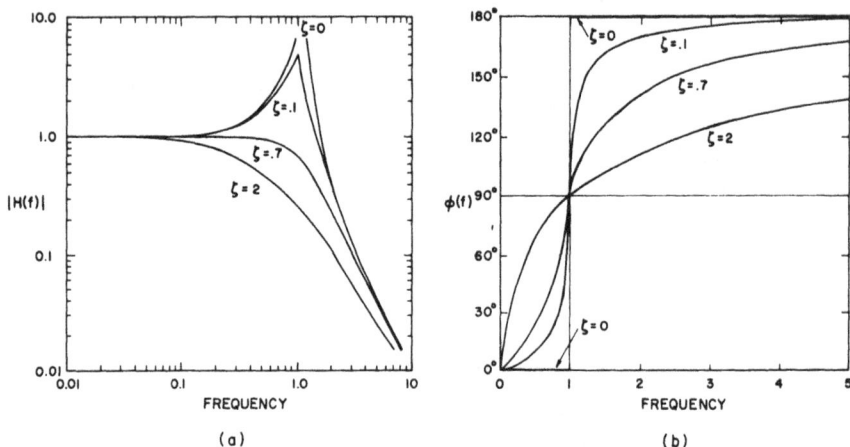

Fig. 3.7.   The frequency response function of an accelerometer; (a) gain factor, (b) phase factor.

Returning to the original purpose of the example, we will calculate the output of the accelerometer. A step function input will be assumed;

$$x(t) = u(t) = \text{unit step function}. \qquad (3.66)$$

The output $V(s)$ is the product of the transform of the input and the transfer function

$$\mathcal{L}[u(t)] = \frac{1}{s}$$

$$V(s) = \left(\frac{1}{s}\right)\left(\frac{-1}{s^2 + 2\zeta\omega_n s + \omega_n^2}\right). \qquad (3.67)$$

To find the output in the time domain, the inverse Laplace transform of the above equation must be calculated. As a first step in the inversion, the poles of the denominator are calculated;

$$D(s) = (s)(s^2 + 2\zeta\omega_n s + \omega_n^2) = 0. \tag{3.68}$$

The quantity $s = 0$ is a root of the first factor, so the first pole v_1 is equal to zero. The poles associated with the second factor can be found from the formula for finding the roots of a quadratic equation, $ax^2 + bx + c$. The discriminant is

$$x = \frac{-b \pm \sqrt{b^2 - 4ac}}{2a}$$

and the roots are

$$v_2 = \omega_n(-\zeta + \sqrt{\zeta^2 - 1})$$
$$v_3 = \omega_n(-\zeta - \sqrt{\zeta^2 - 1}). \tag{3.69}$$

The output can now be expressed in the following partial fraction series expression:

$$V(s) = \frac{b_1}{s} + \frac{b_2}{s - \omega_n(-\zeta + \sqrt{\zeta^2 - 1})} + \frac{b_3}{s - \omega_n(-\zeta - \sqrt{\zeta^2 - 1})}. \tag{3.70}$$

The values of the numerators are calculated next. These are

$$b_1 = \lim_{s \to 0} [sV(s)] = \frac{1}{\omega_n^2}$$

$$b_2 = \lim_{s \to \alpha} [\alpha V(s)], \text{ where } \alpha = \omega_n(-\zeta + \sqrt{\zeta^2 - 1})$$

$$= \frac{1}{2\omega_n^2(-\zeta\sqrt{\zeta^2 - 1} + \zeta^2 - 1)}$$

$$b_3 = \lim_{s \to \beta} [\beta V(s)], \text{ where } \beta = \omega_n(-\zeta - \sqrt{\zeta^2 - 1})$$

$$= \frac{1}{2\omega_n^2(\zeta\sqrt{\zeta^2 - 1} + \zeta^2 - 1)}. \tag{3.71}$$

The inverse transform is found from Eq. (3.70);

$$v(t) = b_1 e^{r_1 t} + b_2 e^{r_2 t} + b_3 e^{r_3 t}$$

$$= \frac{1}{\omega_n^2} + \frac{1}{2\omega_n^2(-\zeta\sqrt{\zeta^2 - 1} + \zeta^2 - 1)} e^{\omega_n(-\zeta + \sqrt{\zeta^2 - 1})t}$$

$$+ \frac{1}{2\omega_n^2(\zeta\sqrt{\zeta^2 - 1} + \zeta^2 - 1)} e^{\omega_n(-\zeta - \sqrt{\zeta^2 - 1})t} \tag{3.72}$$

$$v(t) = \frac{1}{\omega_n^2}\left\{1 - e^{-\zeta\omega_n t}\left[\frac{e^{\omega_n\sqrt{\zeta^2 - 1}\,t}\left(1 + \frac{\zeta}{\sqrt{\zeta^2 - 1}}\right) + e^{-\omega_n\sqrt{\zeta^2 - 1}\,t}\left(1 - \frac{\zeta}{\sqrt{\zeta^2 - 1}}\right)}{2}\right]\right\}.$$

Since $\zeta < 1$ for practical accelerometers, only this case will be solved;

$$v(t) = \frac{1}{\omega_n^2} \left\{ 1 - e^{-\zeta \omega_n t} \left[\left(\frac{e^{j\omega_n \sqrt{1-\zeta^2}\, t} + e^{-j\omega_n \sqrt{1-\zeta^2}\, t}}{2} \right) \right. \right.$$

$$\left. \left. \left(\frac{\zeta}{\sqrt{1-\zeta^2}} \right) \left(\frac{e^{j\omega_n \sqrt{1-\zeta^2}\, t} - e^{-j\omega_n \sqrt{1-\zeta^2}\, t}}{2j} \right) \right] \right\}$$

$$= \frac{1}{\omega_n^2} \left[1 - \left(\frac{e^{-\zeta \omega_n t}}{\sqrt{1-\zeta^2}} \right) \sin \left(\omega_n \sqrt{1-\zeta^2}\, t + \phi \right) \right], \qquad (3.73)$$

where

$$\phi = \tan^{-1} \left[\frac{\zeta}{\sqrt{1-\zeta^2}} \right]. \qquad (3.74)$$

This response is plotted in Fig. 3.8 for several damping ratios. Notice that the output of the accelerometer is not a faithful representation of the actual acceleration. The response oscillates in an exponentially decaying manner about the true step level. As the damping increases, the oscillations decay at a faster rate.

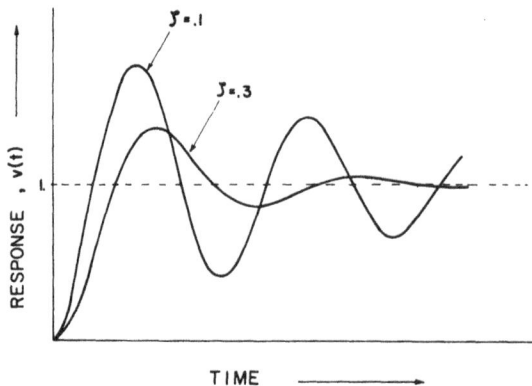

Fig. 3.8. Accelerometer outputs for a step acceleration input.

Response values expressed as Fourier transforms of causal functions ($x(t) = 0$ for $t < 0$) can be converted to Laplace transforms if this step helps to perform the inversion from a frequency function to a time function.

If the Fourier transform is an analytic function of frequency and the Laplace transform exists for the real part of the complex frequency equal to or greater than zero, the Laplace transform is found simply by replacing the real frequency variable f by the quantity $s/(j2\pi)$ (see Ref. 14, p. 173);

$$\mathscr{L}(s) = F\left(\frac{s}{j\,2\pi}\right), \qquad Re[s] \geq 0, \tag{3.75}$$

where $F(f)$ = the Fourier transform.

The restriction on the function existing for zero and positive real values basically means that all poles must be to the left of the $j\omega$ axis in the s plane.

If the Fourier transform is not available as an analytical function, the Laplace transform can be calculated by convolving the Fourier transform of the data and the frequency function $1/(\sigma + j\,2\pi f)$. (This convolution results from expressing the Laplace transform as the Fourier transform of the product of the data and a term containing the unit step function multiplied by $e^{-\sigma t}$. The unit step function forces the integral to exist over positive time only, and the $e^{-\sigma t}$ term changes the kernel of the transformation from a real frequency variable to a complex frequency variable. The Fourier transform of the product of two time functions is the convolution of the separate Fourier transforms. The Fourier transform of $e^{-\sigma t}$ multiplied by step function yields the $1/(\sigma + j\,2\pi f)$ term.)

Thus,

$$\mathscr{L}(s) = \int_{-\infty}^{\infty} \frac{F(u)}{s - ju}\, du \qquad Re[s] \geq 0. \tag{3.76}$$

For causal time functions only, the real $(Re(f))$ or the imaginary $(Im(f))$ portion of the Fourier transform is required [14];

$$\mathscr{L}(s) = 4s \int_{0}^{\infty} \frac{Re(u)}{s^2 + u^2}\, du, \qquad Re[s] \geq 0$$

$$= -4 \int_{0}^{\infty} \frac{u Im(u)}{s^2 + u^2}\, du, \qquad Re[s] \geq 0. \tag{3.77}$$

3.5 Cascaded Linear Systems

When simple linear systems are connected in tandem as shown in Fig. 3.9, the overall frequency response function is the product of the individual frequency response functions as long as the systems do not load each other;

$$H_T(f) = H_1(f) \cdot H_2(f) \cdot H_3(f), \tag{3.78}$$

where $H_T(f)$ = the overall frequency response function.

In terms of the gain and phase factors,

$$|H_T(f)| = |H_1(f)| \cdot |H_2(f)| \cdot |H_3(f)|$$

$$\theta_T(f) = \theta_1(f) + \theta_2(f) + \theta_3(f). \tag{3.79}$$

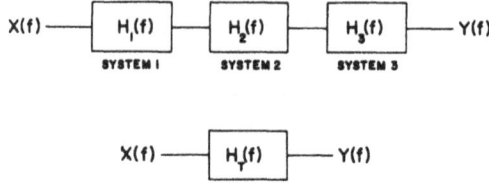

Fig. 3.9. Unloaded simple linear systems in tandem.

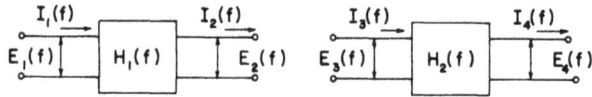

Fig. 3.10. Linear systems in tandem.

If the input of one system loads the output of the system to which it is connected, a different approach must be taken to determine the overall frequency response function. For example, consider the two linear systems shown in Fig. 3.10.

The output of each system can be related in matrix form to its input by means of four pole parameters (see Ref. 15, p. 143);

$$\begin{vmatrix} E_1 \\ I_1 \end{vmatrix} = \begin{vmatrix} A_1 B_1 \\ C_1 D_1 \end{vmatrix} \begin{vmatrix} E_2 \\ I_2 \end{vmatrix} \quad \text{and} \quad \begin{vmatrix} E_3 \\ I_3 \end{vmatrix} = \begin{vmatrix} A_2 B_2 \\ C_2 D_2 \end{vmatrix} \begin{vmatrix} E_4 \\ I_4 \end{vmatrix}, \quad (3.80)$$

where

$$A = \frac{Z_{11}}{Z_{21}} = \frac{y_{22}}{y_{21}}$$

$$B = \frac{[Z]}{Z_{21}} = \frac{1}{y_{21}}$$

$$C = \frac{1}{Z_{21}} = \frac{[y]}{y_{21}}$$

$$D = \frac{Z_{22}}{Z_{21}} = \frac{y_{11}}{y_{21}}$$

$$[Z] = \begin{vmatrix} Z_{11} & Z_{12} \\ Z_{21} & Z_{22} \end{vmatrix}$$

$$Z_{21} = Z_{12} \quad (3.81)$$

$$[y] = \begin{vmatrix} y_{11} & y_{12} \\ y_{21} & y_{22} \end{vmatrix}$$

$$y_{21} = y_{12}.$$

$y_{11} = \dfrac{I_1}{E_1} =$ the input admittance. This is measured when the output of the network is short circuited.

$y_{22} = -\dfrac{I_2}{E_2} =$ the output admittance. This is measured when the input of the network is short circuited.

$y_{12} = -\dfrac{I_1}{E_2} =$ the reciprocal of the transfer impedance. This is measured when the input of the network is short circuited.

$Z_{11} = \dfrac{E_1}{I_1} =$ the input impedance. This is measured with the output open circuited.

$Z_{22} = -\dfrac{E_2}{I_2} =$ the output impedance. This is measured with the input open circuited.

$Z_{12} = -\dfrac{E_1}{I_2} =$ the reciprocal of the transfer admittance. The subscripts on the A, B, C, and D parameters in Eq. (3.80) refer to the two different systems of Fig. 3.10. This is measured with the input open circuited.

Considering one system only, the input is related to the output by the following equations:

$$E_1 = A_1 E_2 + B_1 I_2$$

and $\qquad\qquad\qquad\qquad\qquad\qquad\qquad\qquad\qquad\qquad\quad$ (3.82)

$$I_1 = C_1 E_2 + D_1 I_2.$$

If the system is not loaded, $I_2 = 0$. Defining the frequency response function as the ratio of the output voltage to the input voltage, we have

$$H_1(f) = \frac{E_2(f)}{H_1(f)} = \frac{1}{A_1}. \qquad\qquad (3.83)$$

Now, if the two systems in the preceding figure are connected in tandem, $E_2 = E_3$ and $I_2 = I_3$. Therefore, the input to output relation is

$$\begin{vmatrix} E_1 \\ I_1 \end{vmatrix} = \begin{vmatrix} A_1 B_1 \\ C_1 D_1 \end{vmatrix} \begin{vmatrix} A_2 B_2 \\ C_2 D_2 \end{vmatrix} \begin{vmatrix} E_4 \\ I_4 \end{vmatrix}. \qquad (3.84)$$

This provides a convenient matrix notation for relating any number of systems connected in tandem. For any given system,

$$\begin{vmatrix} E_1 \\ I_1 \end{vmatrix} = \begin{vmatrix} A_1 B_1 \\ C_1 D_1 \end{vmatrix} \cdots \begin{vmatrix} A_n B_n \\ C_n D_n \end{vmatrix} \begin{vmatrix} E_{2n} \\ I_{2n} \end{vmatrix}. \qquad (3.85)$$

From Eq. (3.84), an overall frequency response function for two systems can be calculated between the output voltage of the second system and the input voltage of the first system when they are connected in tandem;

$$\begin{vmatrix} E_1 \\ I_1 \end{vmatrix} = \begin{vmatrix} (A_1 A_2 + B_1 C_2) & (A_1 B_2 + B_1 D_2) \\ (C_1 A_2 + D_1 C_2) & (C_1 B_2 + D_1 D_2) \end{vmatrix} \begin{vmatrix} E_4 \\ I_4 \end{vmatrix}. \qquad (3.86)$$

Since no load is permitted on the second system, $I_4 = 0$ and the overall frequency function $H_T(f)$ is the ratio of $E_4(f)$ to $E_1(f)$;

$$H_T(f) = \frac{E_4(f)}{E_1(f)} = \frac{1}{A_1 A_2 + B_1 C_2}. \tag{3.87}$$

From Eq. (3.87), A_1 and A_2 are the reciprocals of the frequency response functions of systems 1 and 2, respectively. Thus, it can be seen in the general case that to obtain the overall frequency response function when the systems are connected in tandem requires knowledge of the transfer impedance B_1 of the first system and of the transfer admittance C_2 of the second system, in addition to the individual frequency response functions.

Chapter 4

SPECTRAL METHODS

4.1 Spectral Decomposition

The decomposition of a time history is its representation by a combination of simple mathematical functions which can be more easily interpreted by the analyst. Spectral decomposition is used to denote those representations that are functions of frequency. As such, they are the most widely used procedures for the analysis of shock data. The two most common forms of spectral decomposition used in shock data analysis are the Fourier spectrum and the shock spectrum. These two methods are discussed in this chapter.

The popularity of spectral decomposition techniques is based on two primary factors. First, these techniques characterize the data in forms that are related to quantities familiar to all engineers. Steady state impedance values can be used to describe systems. The input and output of any system are described in terms of their frequency content. In the simplest context, these techniques can be thought of as examining a shock to determine if it contains significant inputs at frequencies that are likely to be potentially damaging. Generally, the response of a structure at resonant frequencies will predominate in the total response. Therefore, the content of the shock at these resonant frequencies is of primary concern.

The second factor in favor of spectral techniques is the independence of each frequency component from all others. Thus, it is possible to examine each component separately, ignoring the rest of the spectrum. This allows the computation of the system response to a shock at any specified frequency, provided only that the energy content of the shock and the frequency response function of the system are known at that frequency. For example, to compute the Fourier response spectrum of a linear system at 100 Hz, only the values of the input Fourier spectrum and the frequency response function at 100 Hz must be known (Fig. (4.1)).

This independence in the frequency domain greatly simplifies analyses. By comparison, the calculation of a response in the time domain at some time t_0 is dependent not only on the value of the excitation at that time but on all previous values of the excitation (see Section 3.2, Eq. (3.1)). Finite frequency windows used in actual analyses may cause

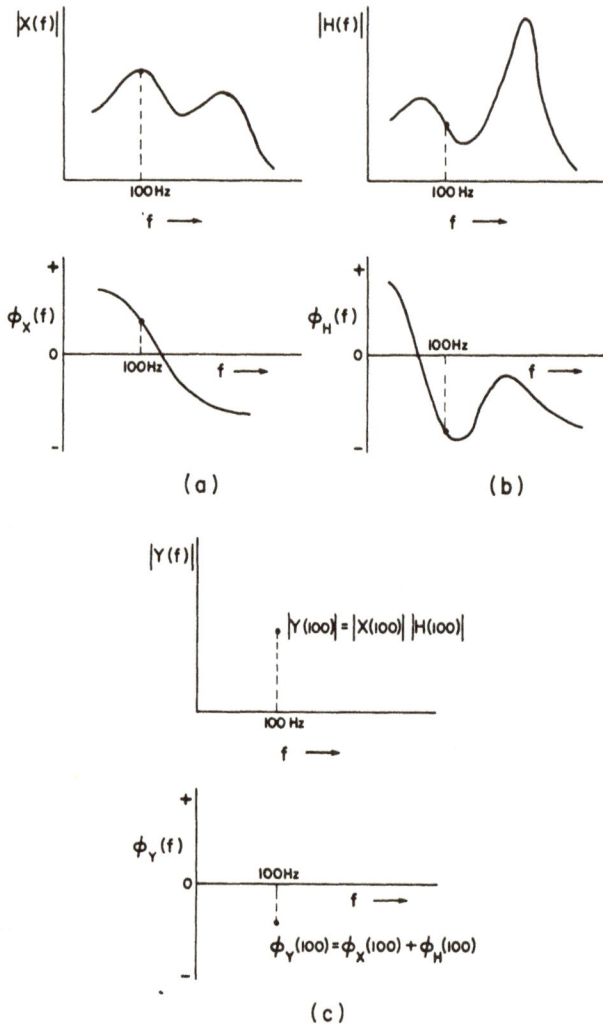

Fig. 4.1. Evaluation of the Fourier spectrum of a response at 100 Hz;
(a) input Fourier spectrum, (b) frequency response function, (c) response
function spectrum.

some dependence between closely spaced spectral components; however,
this dependence is minor compared with that of the time domain.

The Fourier spectrum and the shock spectrum differ significantly.
Fourier analysis consists of decomposing a time history in terms of
trigonometric functions. The magnitude and phase of each trigonometric term needed to reconstruct the time history are plotted as a
function of frequency. Shock spectrum analysis consists of determining the peak response values of a set of simple, second order, mechanical

oscillators to the shock excitation. This peak response is plotted as a function of the undamped natural frequency of the simple oscillators.

Selection between the Fourier and the shock spectral methods of spectral decomposition should be based on the application of the data. Fourier analysis can be used to describe input data, response data, or, if the input and response are measured simultaneously, it can be used to describe the frequency response function of the system. Given any two of the above items, the third can be determined. Typically, the input time history and frequency response function are known, and the response time history is to be determined.

The original, and still the primary, application of shock spectral analysis is to predict peak response levels from input measurements. It is used instead of the Fourier spectrum to predict response levels wherever the frequency response function of the system can be reasonably represented by a simple second order oscillator and when it is not necessary to compute the response in detail. Note that it can only be used with input measurements.

The details of Fourier and shock spectrum analyses are discussed in the following two sections.

4.2 The Fourier Spectrum

The Fourier spectrum is simply the finite, Fourier transform of a time history. This corresponds to evaluating Eq. (4.1a) where the time history $x(t)$ is multiplied by the Fourier kernel, a complex exponential containing both the frequency and time variables, and then integrating this product over the record length;

$$X(f) = \int_{-T/2}^{T/2} x(t) e^{-j2\pi ft} \, dt. \tag{4.1a}$$

Alternatively, the kernel may be expanded so that Eq. (4.1a) may be rewritten as

$$X(f) = \int_{-T/2}^{T/2} x(t) \, \cos 2\pi ft \, dt - j \int_{-T/2}^{T/2} x(t) \, \sin 2\pi ft \, dt. \tag{4.1b}$$

The latter equation shows how the Fourier spectrum can be considered as a decomposition of the time history in terms of sinusoidal and cosinusoidal components.

The Fourier spectrum can be displayed either as real and imaginary functions of frequency or, by means of a complex coordinate transformation, in terms of a modulus and phase angle. The real function corresponds to the first integral in Eq. (4.1b), while the imaginary function is the second integral. The Fourier spectrum is expressed in terms of modulus and phase simply as

$$X(f) = |X(f)| e^{-j\theta(f)}, \tag{4.2}$$

where

$$|X(f)| = \sqrt{\mathrm{Re}^2\ [X(f)] + \mathrm{Im}^2\ [X(f)]} \qquad (4.3)$$

$$\theta(f) = \tan^{-1}\left[\frac{Im\ X(f)}{Re\ X(f)}\right]. \qquad (4.4)$$

Examples of both types of Fourier spectra are shown in Figs. 4.2 and 4.3.

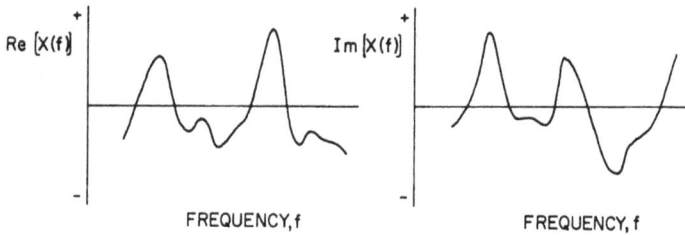

Fig. 4.2.   Fourier spectrum, real and imaginary presentation.

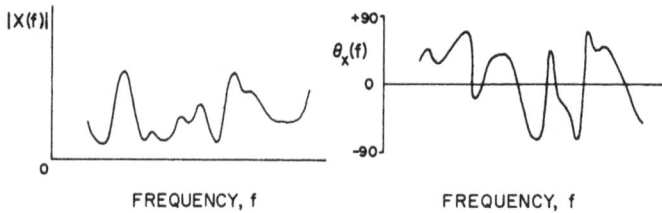

Fig. 4.3.   Fourier spectrum, modulus and phase presentation.

In order to apply Fourier spectrum techniques, the finite Fourier transform of the time history must exist. Necessary and sufficient conditions for the existence of the Fourier transform are discussed in Section 2.3.

An important aspect of the Fourier spectrum is that each of the frequency components is independent of all the others. This is due to the orthogonality of the sine and cosine functions. Strictly speaking, the components will be independent only for infinite record lengths. In this case, the Fourier spectrum is a continuous function of frequency. However, as was shown in Section 2.3, truncating the time history in effect multiplies it by a boxcar weighting function. This is equivalent to convolving the true Fourier spectrum with a sin x/x window. Since the window has finite bandwidth, the finite Fourier spectrum may be thought of as a line spectrum, where the components are spaced at frequency intervals corresponding to the sin x/x window as

$$Y(f) = T[\sin \pi f T / \pi f T],$$

where T is the record length of the time history. From this equation, the half-power points of the main lobe occur at $f = \pm 0.442/T$ with respect to the spectral component over which the sin x/x function is centered. The main lobe zero crossings occur at $f = \pm 1/T$. Utilizing the half-power bandwidth of the sin x/x window as an indication of the frequency spacing of finite spectral components implies that these components occur every $0.884/T$ Hz. The rule of thumb generally used is

$$\Delta f = 1/T, \tag{4.5}$$

where Δf is the frequency interval.

Spectral components may be calculated at finer spacings, but it is important to realize that such components are simply interpolated values obtainable from the *true* components by means of a sin x/x interpolation function. Figure 4.4 indicates the spacing of the sin x/x windows when a Δf of $1/T$ is used.

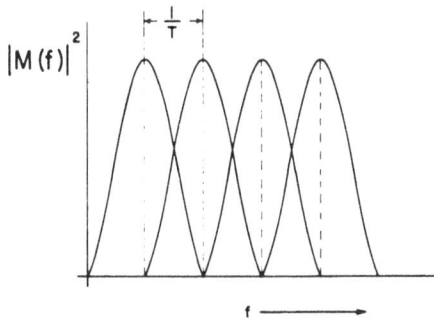

Fig. 4.4.   Spectral windows for 1/T spacing.

For transient data, the maximum record length is defined by the time required for the signal amplitude to decay completely. Extending the record length beyond this point does not reduce the spacing between the spectral components required to completely define the spectrum.

The Fourier spectrum may be computed for either the excitation or the response time history. When dealing with shock data, the difficulties inherent in measuring the response of the system under service conditions frequently limit the analysis to that of the excitation time history. In general, the Fourier spectrum of the shock input is used, along with available knowledge of the frequency response function of the system subjected to the shock, in order to obtain an indication of the system response. Since this response is calculated in the frequency domain, the inverse Fourier transformation operation will produce the response time history of the system. This operation is

$$y(t) = \int_{-\infty}^{\infty} H(f) \, X(f) e^{j2\pi ft} \, df, \qquad (4.6)$$

where $H(f)$ is the frequency response function of the system. If the response of a system is to be computed from the excitation, certain requirements are also placed on the system. Basically, it must conform to the ideal linear system. The system must be stable, physically realizable, and describable by constant-parameter differential equations. Finally, the system must be excited by only one input and must produce only one response. In practice, only the stability and physical realizability can be perfectly met. All physical systems will deviate somewhat from the other requirements. The engineering problem boils down to assuring that these deviations are minor if the Fourier techniques are to be used. Low-level noise, slight nonlinearities, and slow parameter drifts usually can be tolerated. However, they degrade the accuracy of the computations.

Minimal use has been made of Fourier spectrum procedures in the specification of shock tests. However, it is possible to generate a single excitation via Fourier techniques that in some manner represents the service environment of a system as defined by a set of excitation time histories. This is because the average spectrum of a set of time histories is equal to the Fourier spectrum of the average time history. This may be seen from the following relations.

Given a set of time histories x_i, $i = 1, \ldots, m$, then the average time history $\bar{x}(t)$ is defined by

$$\bar{x}(t) = \frac{1}{m} \sum_{i=1}^{m} x_i(t). \qquad (4.7)$$

The Fourier transform of $\bar{x}(t)$ is then

$$\overline{X}(f) = \int_{-\infty}^{\infty} x(t) e^{-j2\pi ft} dt. \qquad (4.8)$$

Replacing $\bar{x}(t)$ in Eq. (4.8) by the expression in Eq. (4.7), we have

$$\overline{X}(f) = \int_{-\infty}^{\infty} \left[\frac{1}{m} \sum_{i=1}^{m} x_i(t) \right] e^{-j2\pi ft} dt. \qquad (4.9)$$

Reordering the integral and summation yields

$$\overline{X}(f) = \frac{1}{m} \sum_{i=1}^{m} \left[\int_{-\infty}^{\infty} x_i(t) e^{-j2\pi ft} dt \right], \qquad (4.10)$$

which is simply another way of writing

$$\overline{X}(f) = \frac{1}{m} \sum_{i=1}^{m} X_i(f). \qquad (4.11)$$

Therefore, the procedure required to determine an estimate of the average frequency content of a transient is quite simple.

For the test of providing some degree of confidence that it produces responses in excess of those caused by the average service environment, the average Fourier spectrum must be increased by some safety factor. Note that if the frequency response function of the system is known, then there is no need to compute an average spectrum. Instead, the responses may be calculated for each measured shock, and the most severe response detected. The shock causing this response may then be used as the test specification.

Determining meaningful safety factors is an area requiring considerable future investigation. The usual approach taken is to estimate the probability density function of the process and also to compute its standard deviation. The required test environment may then be defined as

$$\overline{Y}(f) = \overline{X}(f) + ks(f), \tag{4.12}$$

where

$$s^2(f) = \frac{1}{m-1} \sum_{i=1}^{m} [X_i(f) - \overline{X}(f)]^2, \tag{4.13}$$

and k is the number of standard deviations required to ensure that a large majority of the probable service environments are encompassed. The scale factor k must be obtained from the estimated probability density function. Unfortunately, since the Fourier spectrum is complex, it is difficult to perform the test specification generation as defined by Eqs. (4.12) and (4.13). Other conceivable approaches which consider the two components of the Fourier spectrum independently are

● Enveloping the moduli of the measured shocks and utilizing the average phase factor, and
● Adjusting the average modulus upward by the product of a scale factor and the standard deviation of the spectrum and utilizing the average phase factor.

One other approach is to compute the test specification in the time domain and then calculate its Fourier spectrum. This can be done by calculating the average time history (which is equivalent to computing the average Fourier spectrum) and the time-varying standard deviation. Then the test specification time history is just

$$y(t) = \bar{x}(t) + ks(t), \tag{4.14}$$

where $y(t)$ is the specification time history, $\bar{x}(t)$ is the average time history, $s(t)$ is the standard deviation, and k is the scale factor.

The last method appears to be the most attractive because the calculations are performed with real variables. Also, if the shock can be considered to be primarily deterministic, $s(t)$ is an indication of the extraneous "noise" in the measurements.

Presently, a majority of the testing procedures based on the Fourier spectrum are concerned with the duplication of measured time histories [16]. The decomposition of the time history is performed by a filtering process in conjunction with the equalization of the test equipment. In effect, the equalization procedure divides the Fourier spectrum of the time history by the frequency response function of the test equipment. This guarantees that the excitation produced by the test equipment subjects the test specimen to an environment which closely approximates the measured excitation.

4.3   The Shock Spectrum

The shock spectrum is a method originated by Biot [17] as a means for estimating the damaging effects of seismic shocks upon buildings and has since been used in analyzing shocks which have been applied to a linear system.

When an acceleration is applied to the base of a simple mechanical oscillator of the type shown in Fig. 4.5 the equation of motion of the mass m is as follows:

$$-m\ddot{y}(t) - k[y(t) - x(t)] = 0. \tag{4.15}$$

The first term is the inertial force opposing motion. The second term is the force due to the compression of the spring.

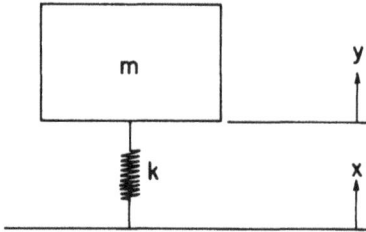

Fig. 4.5.   Base-excited simple mechanical oscillator.

In this system, failure will occur if the stress in the spring exceeds some fixed value. Typically, this may be the yield strength if permanent deformation is the failure criterion, or ultimate strength if rupture of the spring is the failure criterion. The stress in the spring element is the second term in Eq. (4.15). By rearranging this equation as

$$m\ddot{y}(t) = -k[y(t) - x(t)]$$
$$= -k\xi(t), \tag{4.16}$$

where $\xi(t) = y(t) - x(t) =$ relative displacement, it can be seen that for any shock the stress can be determined in either of two ways.

First, the displacement of the mass of the system relative to the base of the system can be calculated and then scaled by the spring constant to determine the stress. Second, the absolute acceleration of the mass of the system can be determined and then scaled by the mass of the system to yield the stress in the spring.

Furthermore, since the stress level is being compared to some fixed value that equates to failure, only the peak value of stress needs to be determined. If the calculated maximum value of stress from a given shock is below the failure criterion value, the system will survive and vice versa, if the calculated value exceeds the criterion value, the system will fail when exposed to the given shock.

In this manner Biot developed a simple method for evaluating the damage potential of very complicated shock time histories on structures. The restrictions on application of this technique are as follows:

1. Peak stress in the spring element is indicative of failure.

2. The entire system can reasonably be represented by the simple mechanical oscillator of Fig. 4.5. (Actually, the concept has been extended to include a viscous damping element between the mass and the base of the system, as discussed later in this chapter. The concept has also been extended to cover certain systems more complicated than the above one. This discussion is in Chapter 7.)

3. The shock spectrum of the excitation exists. (The necessary conditions for the existence of a shock spectrum have not been derived; however, sufficient conditions should be met by any continuous signal of finite energy and duration.)

The shock spectrum is defined as the maximum response of a set of linear second order systems to the shock recorded as a function of the natural frequency of these systems. This corresponds to playing the shock time history through a series of systems whose frequency response functions are of the type shown in Fig. 4.6 and plotting the peak response of each system. Figure 4.7 is a typical shock spectrum.

To use the shock spectrum for failure analysis, the following four pieces of information are required:

1. The undamped natural frequency of the system
2. The minimum stress level which will cause failure
3. The mass of the system
4. The cross-sectional area to which the shock is applied.

As an example, suppose that a system with a natural frequency of 100 Hz, a maximum allowable stress level of 5000 lb/sq in., a mass of 5 lb-sec^2/ft, and a surface area of 2 sq in. is to be exposed to the transient that produced the above shock spectrum. The peak acceleration response is read from the shock spectrum at 100 Hz. This is 10 g's. The maximum stress is then the product of the peak acceleration and the system's mass divided by the surface area,

$$10 \text{ (g's)} \times 32.2 \text{ (ft/sec}^2 - \text{g)} \times 5 \text{ (16-sec}^2\text{/ft)}/2 \text{ (in.}^2) = 805 \text{ (lb/in.}^2).$$

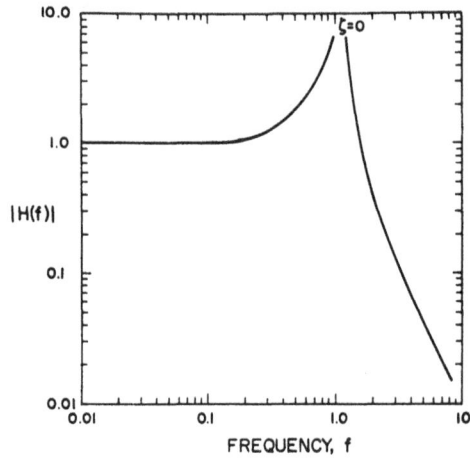

Fig. 4.6. The frequency response function
of a simple mechanical oscillator.

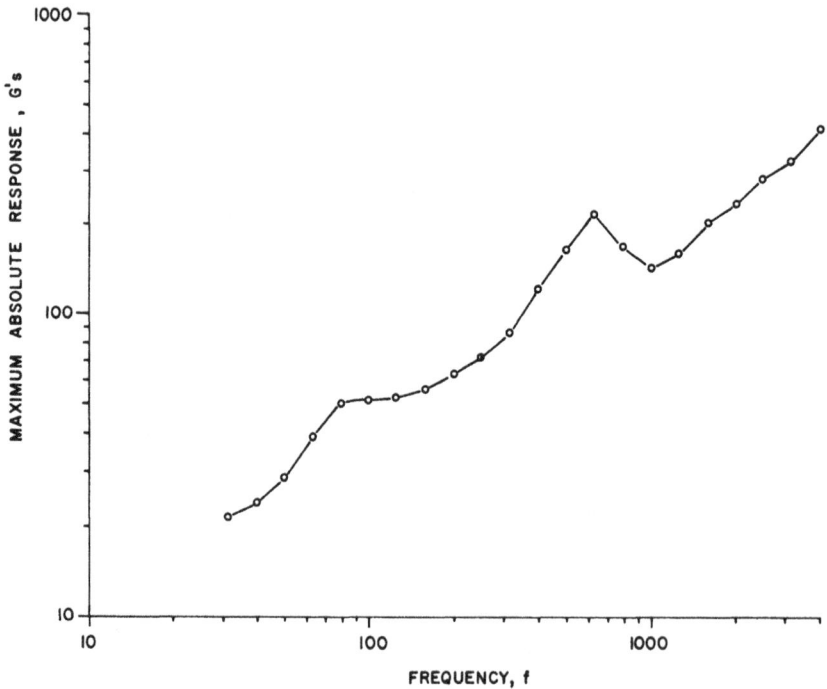

Fig. 4.7. A typical shock spectrum.

Since this is below the allowable stress level, the system will not fail when exposed to this shock.

The definition of the shock spectrum is actually much broader than it appears at first glance. To reduce the ambiguities in this definition, the clarification of several items is required. These items are

- Type of excitation
- Type of response
- Point of excitation application
- Type of spectrum
- Damping.

Each of these will now be discussed in turn.

Type of Excitation

The excitation may consist either of a force or a motion applied to the system. Motion may be measured in terms of deflection, velocity, or acceleration imparted to the system by the shock. In most cases, an accelerometer is used to transduce the excitation so that most shock spectrum procedures assume acceleration inputs.

Type of Response

The next item concerns the type of response required. Two different types of classification are necessary here. First of all, the response of the mass of the system may be referenced relative either to the base of the system or to some fixed point in space. The first of these is termed *relative* motion, while the second is *absolute*. A second classification concerns the type of response motion. This motion may be defined in terms of its deflection, velocity, or acceleration. The combination of these two classifications leads to six different responses which fit the shock spectrum definition. These are

1. Relative deflection $\xi(t)$
2. Relative velocity $\dot{\xi}(t)$
3. Relative acceleration $\ddot{\xi}(t)$
4. Absolute deflection $y(t)$
5. Absolute velocity $\dot{y}(t)$
6. Absolute acceleration $\ddot{y}(t)$.

In actual practice, two other responses are also used. The first of these is the pseudovelocity $V(t)$ defined by

$$V(t) = 2\pi f_n \xi(t), \qquad (4.17)$$

where f_n is the natural frequency of the system. The pseudovelocity is identical with the relative velocity when the system response is a pure sinusoid. This is true only for the residual response of an undamped single degree-of-freedom system. In general, the two will be in close agreement, with the pseudovelocity lower at low frequencies and higher at high frequencies than the true relative velocity because of its frequency dependence.

The other response is called the *equivalent static acceleration* $A_{eq}(t)$, which corresponds to the true absolute acceleration only for an undamped system. It is defined by

$$A_{eq}(t) = 2\pi f_n V(t) = 4\pi^2 f_n^2 \xi(t).$$ (4.18)

Because of its frequency dependence, it also will tend to be lower at low frequencies than the true absolute acceleration. However, it will be in good agreement with the true absolute acceleration for high frequencies.

Since these two quantities are easily derived from the relative deflection, they are often presented simultaneously with $\xi_{max}(f)$ as a function of frequency by means of a four-coordinate nomographic grid as shown in Fig. 4.8. The two response motions most often used in shock spectrum analysis are relative deflection and absolute acceleration.

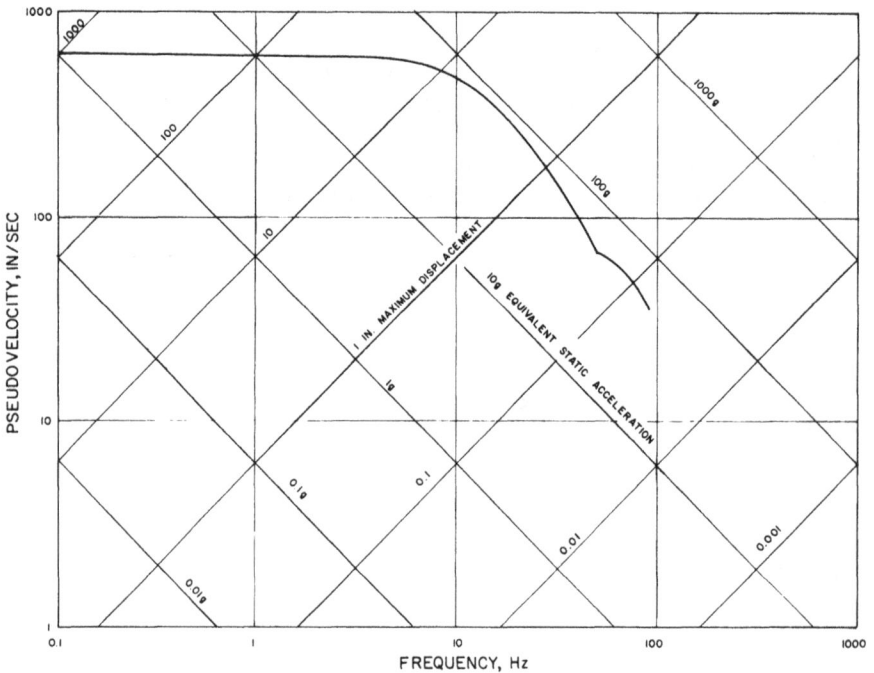

Fig. 4.8.   A four-coordinate shock spectrum.

Point of Excitation Application

The next consideration is the part of the system to be excited. The excitation may be applied either to the base as shown in Fig. 4.9a or directly to the mass as shown in Fig 4.9b. The decision as to the application point should be made by modeling the physical system being analyzed. The base-excited system is the one chosen in most cases.

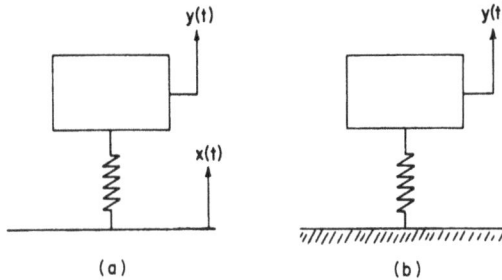

Fig. 4.9.   Base-input (a) and mass-input (b) versions of the simple mechanical oscillator.

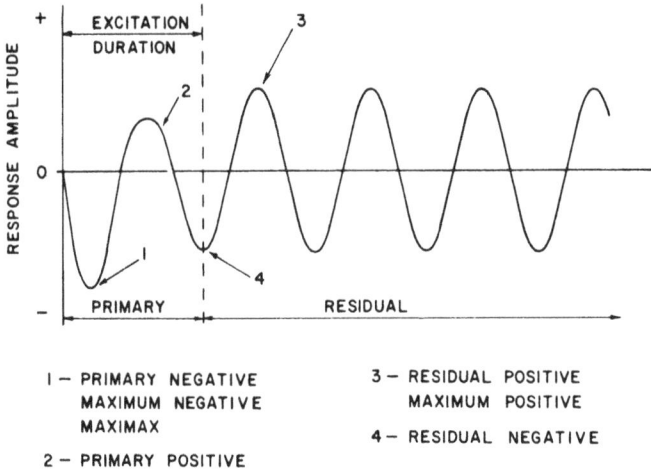

Fig. 4.10.   Response maxima.

Type of Spectrum

The next item to be clarified is the type of spectrum. At least seven different types of spectra are presently in use. These values for one frequency are illustrated in the response time history shown in Fig. 4.10. They may be classified by the polarity of the peak response and its time of occurrence. The *maximax* spectrum consists of the maximum absolute response recorded as a function of the system natural frequency. The *maximum positive* spectrum contains only the maximum positive response excursions, while the *maximum negative* spectrum contains only the maximum negative excursions. The *primary* spectrum is made up of the maximum absolute responses during the excitation, while the *residual* spectrum contains the peak response occurring after the excitation has completely decayed. Primary positive, primary negative, residual positive, and residual negative spectra are also used. Their definitions are obvious and follow from their names. Note that the maximax spectrum envelopes all of the other types and as a result is the one most often used. However, the selection of the proper type should be governed by the application of the spectrum.

Damping Considerations

Finally, the shock spectrum may be damped or undamped. The classical definition of the method assumed no damping because the intent was to obtain a conservative estimate of the damage potential of the shock. Since it is usually quite difficult to determine the *true* damping of the system (in fact, the damping will vary with the particular mode of the system which has been excited by the shock) present procedures consist of calculating the shock response at several different critical damping ratio values in the range 0 to 0.1 (from undamped to 10 percent of critical damping).

The primary purpose for including damping is to reduce the over-conservatism of shock spectral analyses, particularly with shocks that contain strong oscillatory components. The effects of including damping are to negate the exact equivalence between

1. The inertial force and the spring force
2. The relative velocity and the pseudovelocity
3. The absolute acceleration and the equivalent static acceleration.

When damping is included as shown in Fig. 4.11, the equation of motion is

$$-m\ddot{y}(t) - c\dot{\xi}(t) - k\xi(t) = 0. \tag{4.19}$$

Thus the stress (spring force) is equal to the sum of the inertial and the damping forces.

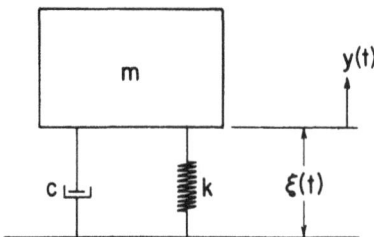

Fig. 4.11. A simple mechanical oscillator with damping.

Pseudovelocity and equivalent static acceleration are no longer equal to the true relative velocity and absolute acceleration because the true relative displacement is no longer sinusoidal, so that integration and double integration do *not* correspond to multiplication by $2\pi f_n$ and $(2\pi f_n)^2$, respectively. Damping also causes the relative displacement response to decay from the pure sine wave of the undamped case.

Two Basic Types of Shock Spectra

Further discussions of the shock spectrum will be limited to the two basic types in widest use. These are

1. The maximax spectrum derived from a base-excited system with absolute acceleration input and relative deflection response

2. The maximax spectrum derived from a base-excited system with absolute acceleration input and absolute acceleration response.

For the first case, the differential equation of the system is written in terms of $\xi(t) = y(t) - x(t)$:

$$\ddot{\xi}(t) + 2\zeta\omega_n\dot{\xi}(t) + \omega_n^2\xi(t) = -\ddot{x}(t), \qquad (4.20)$$

where $\omega_n = 2\pi f_n$, ζ is the critical viscous damping ratio, and $\ddot{x}(t)$ is the absolute acceleration of the excitation. The general solution is

$$\xi(t) = \xi_0 e^{-\zeta\omega_n t}\cos\omega_n t + \frac{\zeta}{\sqrt{1-\zeta^2}}\sin\omega_d t + \frac{\dot{\xi}_0}{\omega_d}e^{-\zeta\omega_n t}\sin\omega_d t$$

$$-\frac{1}{\omega_d}\int_0^t \ddot{x}(\tau)e^{-\zeta\omega_n(t-\tau)}\sin\omega_d(t-\tau)d\tau, \qquad (4.21)$$

where $\xi_0 = \xi(t_0)$ and $\dot{\xi}_0 = \dot{\xi}(t_0)$ are the system's relative deflection and velocity just prior to the shock and ω_d is the damped natural frequency defined by

$$\omega_d = \omega_n\sqrt{1-\zeta^2}. \qquad (4.22)$$

Note that for zero initial conditions,

$$\xi(t) = -\frac{1}{\omega_d}\int_0^t \ddot{x}(\tau)e^{-\zeta\omega_n(t-\tau)}\sin\omega_d(t-\tau)d\tau. \qquad (4.23)$$

This may be recognized as the convolution of $\ddot{x}(t)$ with the unit impulse response of the single degree-of-freedom system. This convolution is called the Duhamel or superposition integral.

The term superposition is descriptive of the fact that the excitation may be viewed as a series of impulses of duration $\Delta\tau$ as shown in Fig. 4.12, and the total system response consists of the superimposed responses to these impulses.

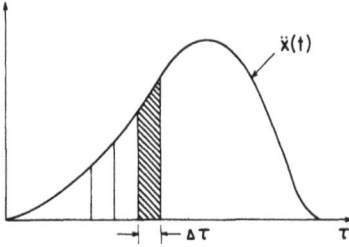

Fig. 4.12.   Excitation viewed as a series of impulses.

The differential equation for the absolute acceleration response is written in terms of the absolute deflection $y(t)$,

$$\ddot{y}(t) + 2\zeta\omega_n\dot{y}(t) + \omega_n^2 y(t) = 2\zeta\omega_n\dot{x}(t) + \omega_n^2 x(t). \tag{4.24}$$

The solution, in terms of the absolute acceleration $\ddot{y}(t)$, is just

$$\ddot{y}(t) = 2\zeta\omega_n[\dot{y}(t) - \dot{x}(t)] - \omega_n^2[y(t) - x(t)] \tag{4.25}$$

or

$$\ddot{y}(t) = -2\zeta\omega_n\dot{\xi}(t) - \omega_n^2\xi(t), \tag{4.26}$$

where $\dot{\xi}(t)$ may be obtained by differentiating Eq. (4.21). This yields

$$\dot{\xi}(t) = -\xi_0\omega_n e^{-\zeta\omega_n t}\sin\omega_d t + \dot{\xi}_0 e^{-\zeta\omega_n t}\left(\cos\omega_d t - \frac{\zeta}{\sqrt{1-\zeta^2}}\sin\omega_d t\right)$$

$$- \int_0^t \ddot{x}(\tau)e^{-\zeta\omega_n(t-\tau)}\left[\cos\omega_d(t-\tau) - \frac{\zeta}{\sqrt{1-\zeta^2}}\sin\omega_d(t-\tau)\right]d\tau. \tag{4.27}$$

For zero initial conditions,

$$\dot{\xi}(t) = -\int_0^t x(\tau)e^{-\zeta\omega_n(t-\tau)}\cos\omega_d(t-\tau)\,d\tau$$

$$+ \frac{\zeta}{\sqrt{1-\zeta^2}}\int_0^t \ddot{x}(\tau)e^{-\zeta\omega_n(t-\tau)}\sin\omega_d(t-\tau)\,d\tau \tag{4.28}$$

or

$$\dot{\xi}(t) = -\int_0^t x(\tau)e^{-\zeta\omega_n(t-\tau)}\cos\omega_d(t-\tau)\,d\tau - \zeta\omega_n\xi(t). \tag{4.29}$$

In the usual case where the system is at rest prior to the shock,

$$\ddot{y}(t) = 2\zeta\omega_n\int_0^t \ddot{x}(\tau)e^{-\zeta\omega_n(t-\tau)}\cos\omega_d(t-\tau)\,d\tau + (2\zeta^2-1)\omega_n^2\xi(t). \tag{4.30}$$

Techniques for implementing these equations on analog and digital equipment are discussed in Chapters 5 and 6.

Relationships Between Shock and Fourier Spectra

Although the shock spectrum and Fourier spectrum are completely different concepts, there exists a relationship between them. The modulus of the Fourier spectrum is identical to the residual, undamped shock spectrum when it is computed in terms of pseudovelocity. This fact is pointed out in Ref. 18. In Ref. 19 this approach is used to compute both the shock spectrum and the Fourier spectrum in one operation. This relationship can be seen by examining Eqs. (4.23) and (4.29). The response at the instant of shock termination (assuming zero initial conditions) is

$$\xi(t) = -\frac{1}{\omega_n} \int_0^t \ddot{x}(\tau) \, \sin \omega_n(t-\tau) d\tau. \tag{4.31}$$

The velocity response is

$$\dot{\xi}(t) = -\int_0^t \ddot{x}(\tau) \, \cos \omega_n(t-\tau) d\tau. \tag{4.32}$$

Expanding the $\sin \omega_n(t-\tau)$ and $\cos \omega_n(t-\tau)$, and then expressing $\ddot{x}(t)$ in terms of its Fourier spectrum at the natural frequency of the system result in

$$\omega_n\xi(t) = [Re\,X(\omega_n)] \, \sin \omega_n t + [Im\,X(\omega_n)] \, \cos \omega_n t \tag{4.33}$$

and

$$\dot{\xi}(t) = [Re\,X(\omega_n)] \, \cos \omega_n t - [Im\,X(\omega_n)] \, \sin \omega_n t. \tag{4.34}$$

Since the system is undamped, the residual response will be a sine wave with $\xi(t)$ and $\dot{\xi}(t)$ as initial conditions. The maximum value of this residual response is just the amplitude of this sine wave, which may be expressed as

$$(\xi_r)_{max} = \sqrt{\xi^2(t) + \frac{\dot{\xi}^2(t)}{\omega_n^2}}. \tag{4.35}$$

Substituting the relationships of Eqs. (4.33) and (4.34) into Eq. (4.35) yields

$$\omega_n(\xi_r)_{max} = \sqrt{[Re\,X(\omega_n)]^2 + [Im\,X(\omega_n)]^2}. \tag{4.36}$$

Since the left side of Eq. (4.36) is just the residual shock spectrum at frequency ω_n expressed in terms of pseudovelocity, and the right side is the modulus of the Fourier spectrum at the same frequency, it follows that

$$(V_r)_{max} = |X(\omega)|. \tag{4.37}$$

Applications of Shock Spectrum Analysis

The major advantages of shock spectrum analysis over other techniques are that neither the detailed frequency response function nor the detailed time history of the system's response to the shock need be known. In many instances it is either difficult or inconvenient to measure these quantities. In some cases, just the effect of mounting a transducer on the system for the purpose of measuring its response is sufficient to change that response significantly.

The shock spectrum is used to determine whether a system can survive a shock environment, to design the system to survive shock environments, and to specify shock tests. The use of the shock spectrum to derive shock test specifications has several advantages. First, since the shock spectrum contains more information about the shock than just its peak value, it is a more accurate test than the simple tests such as dropping the system from a given height onto a specified surface. In particular, it provides much greater repeatability. Second, since the test is derived from measurements of the actual service environment, it more closely simulates the damaging effects of the service environment than the simple tests, even those with simple repeatable pulse shapes. Third, it has generally been accepted that, since a conservative test is desired, a specification can be created by simply enveloping all the measured shock spectra of a particular shock environment. This is attractive because of its simplicity.

On the other hand, there are several disadvantages to using the shock spectrum to develop shock test specifications. First, there is no unique relationship between a shock spectrum and a time history, as there is with a Fourier spectrum. Many time histories can produce the same shock spectrum. This fact has been used to implement some shock testing with vibration test equipment [20]. Exponentially decaying sinusoids have been summed to provide test inputs, and sinusoidal sweeps, with the amplitude programmed as a function of frequency, have also been used as shock test inputs. In both cases, the response spectrum is duplicated.

This many-to-one relationship between the excitation and the shock spectrum is also the cause of considerable controversy over the validity of the shock spectrum as a measure of damage potential. It has been argued that it is not possible to determine true damage potential in this manner with any degree of certainty because each of the various exciting pulses which could produce a given shock spectrum will have a different damage potential associated with it. This is highly dependent on the failure mechanism in the system. If the single highest peak criterion applies to the system failure mode, then duplication of the response spectrum is adequate. However, many systems fail in different manners. For example, if the primary failure mode is from fatigue, then any testing method that causes more cycles of high-level responses than the actual environment will be an overly conservative test.

A second disadvantage of the present use of shock spectra for specification purposes is that the enveloping of peak values, while easy to implement, also probably results in quite conservative tests. A more realistic approach would be to treat a group of shock spectra from the same nominal environment as random data. Then by ensemble averaging, the mean spectrum and its variance could be computed. The test specification could then be based on the mean increased by the product of some factor and the standard deviation of the group of shock spectra. When combining spectra, care must be taken to assure that the spectra are equivalent. For example, if there have been changes in the system between measurements, it may be necessary to modify some of the time histories (see Section 3.5) and to recompute the shock spectra before combining. See Ref. 21.

A third problem is the difficulty in determining the accuracy of the shock spectrum, since implicit in the derivation is the assumption that the physical system being tested may be approximated by a single degree-of-freedom system. Regardless of these difficulties, the shock spectrum is employed extensively in the analysis of shock data.

In recent years, the shock spectrum concept has been extended to multiple degree-of-freedom system, nonlinear systems, etc. These extensions are discussed in Chapter 7.

Chapter 5

ANALOG TECHNIQUES FOR ANALYZING SHOCK DATA

Analog equipment used for the analysis of shock data can be separated into two categories. These are the special purpose and the general purpose equipment categories. Special purpose equipment is that designed primarily for the analysis of transient data, while the general purpose equipment, of course, has many other applications.

Each equipment category can be further divided into two subgroups. These subgroups designate the underlying manner in which the equipment operates. The equipment in these two subgroups are known either as mathematical or physical analog computers.

This chapter describes the basic operating principles of each of these analog computation techniques and in particular their application to the analysis of shock data.

5.1 Electrical Analogs

At the present state of the art, dynamic physical quantities are normally measured by a transducer that has an electrical output. The electrical output from the measurement system is almost always a voltage. This voltage will bear a known relationship to the physical quantity of interest.

Since data are normally acquired as a voltage, it is reasonable that the equipment used to analyze the data work directly on a voltage signal. There are, of course, many other advantages to working on the electrical signal. Interestingly enough, some of the early shock analyzers were mechanical [1, 22]. The first was a torsional pendulum whose resonant frequency was varied by changing the position of the weight on the pendulum. The second consisted of a bank of reed gages tuned to different frequencies. The maximum response of each reed was measured to obtain the shock spectrum. Galvanometers have also been used to form an electromechanical shock analyzer [23]. In this chapter, only purely electrical analyzers will be considered.

As noted before, analog computing circuits are of two varieties. In the mathematical analog variety, the equipment implements the mathematical operations required to solve the problem. General purpose analog computers of this type are known as electronic differential analyzers. Analog computers of the physical analog variety are usually known as passive analog computers.

Electronic Differential Analyzers

The usual application of electronic differential analyzers is to the solution of differential equations. The mathematical operations that are most convenient to implement on the computer are
- Addition
- Integration
- Sign change (polarity change), and
- Scale factoring (multiplication by a constant).

These operations are generally adequate to solve linear, constant-coefficient differential equations.

Implementation of nonlinear operations, including multiplication and division by time-varying quantities, can be accomplished although over a more restricted range and usually with less accuracy than the above-listed functions. Differentiation is implemented only when there is no alternative route for solving the problem (the differentiation operation causes noise and instability problems).

The manner in which the mathematical analog computer operates is as follows:

1. The highest order derivative is found as a function of all of the other parameters in the system equation.

2. A voltage proportional to this highest order derivative is assumed to exist. It is integrated an appropriate number of times to obtain all of the other terms in the equation, except the input term.

3. These other terms and the input term are added together (with the appropriate signs and scale factors) to form the voltage proportional to the highest order derivative.

4. The voltage corresponding to the output term of interest is read out.

As an illustration of the use of an electronic differential analyzer, the circuit will be developed to solve for the absolute acceleration of the mass of a simple mechanical oscillator when a base acceleration is applied.

The mechanical system is shown in Fig. 5.1.

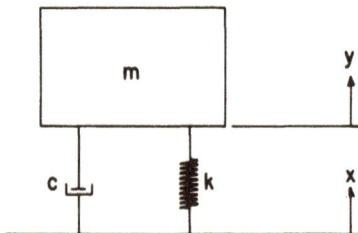

Fig. 5.1. Base-excited simple mechanical oscillator.

The motion of this system is described by

$$-\ddot{y}(t) - 2\zeta\omega_n[\dot{y}(t) - \dot{x}(t)] - \omega_n^2[y(t) - x(t)] = 0, \tag{5.1}$$

where

$x(t) =$ the displacement of the base

$\dot{x}(t) =$ the velocity of the base

$\ddot{x}(t) =$ the acceleration of the base

$y(t) =$ the displacement of the inertial mass

$\dot{y}(t) =$ the velocity of the inertial mass

$\ddot{y}(t) =$ the acceleration of the inertial mass

$$\zeta = \frac{c}{2\sqrt{km}} = \text{the fraction of critical damping}$$
of the mechanical oscillator

$$\omega_n = \sqrt{\frac{k}{m}} = \text{the undamped natural frequency}$$
of the mechanical oscillator.

The highest order derivative is found as follows:

$$\ddot{y}(t) = 2\zeta\omega_n[\dot{x}(t) - \dot{y}(t)] + \omega_n^2[x(t) - y(t)]. \tag{5.2}$$

A voltage equal to the acceleration of the mass is assumed to exist, and it is integrated once to obtain the velocity of the mass. Then this velocity is integrated to obtain the displacement of the mass. The input is a voltage representing base acceleration. This must be integrated once to obtain the base velocity term on the right-hand side of Eq. (5.2), and a second time to obtain the base displacement term. The two terms are subtracted as are the displacement terms. The velocity difference term is multiplied by the constant $2\zeta\omega_n$ and then these two factors are summed to yield the acceleration of the mass. The schematic diagram for an analog computing circuit to solve Eq. (5.1) is shown in Fig. 5.2. Each integrating and summing amplifier has 180° of phase shift so that there is a polarity change between their inputs and outputs.

The symbols used in Fig. 5.2 are as follows:

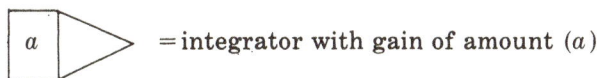

a = integrator with gain of amount (a)

a = summing amplifier with gain of amount (a)

The circuit of Fig. 5.2 is not optimum. It is primarily intended to illustrate the method of creating the analog circuit. One method of simplify-

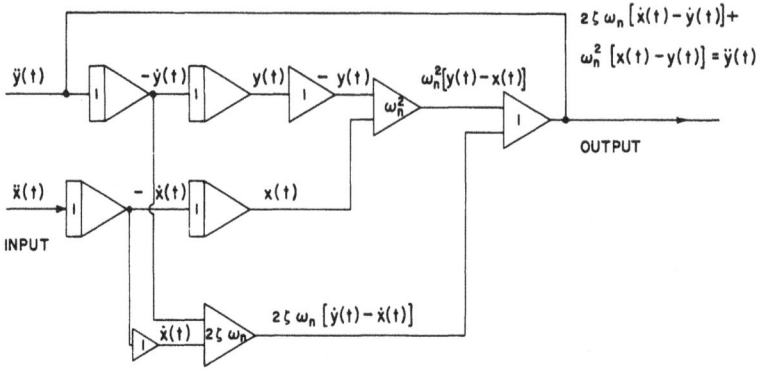

Fig. 5.2. Analog computer circuit for determining the mass acceleration response of a simple mechanical oscillator to a base acceleration input (for explanation of symbols, see p. 83).

ing the preceding circuit to reduce the total number of integrators and summing amplifiers needed is to assume that a voltage proportional to the relative acceleration of the mass is available. Let

$$\ddot{\xi}(t) = \ddot{y}(t) - \ddot{x}(t) = \text{the acceleration of the inertial mass of the simple mechanical oscillator relative to the base}$$

$$\dot{\xi}(t) = \dot{y}(t) - \dot{x}(t) = \text{the velocity of the inertial mass of the simple mechanical oscillator relative to the base}$$

$$\xi(t) = y(t) - x(t) = \text{the displacement of the inertial mass of the simple mechanical oscillator relative to the base.}$$

The circuit as simplified is shown in Fig. 5.3.

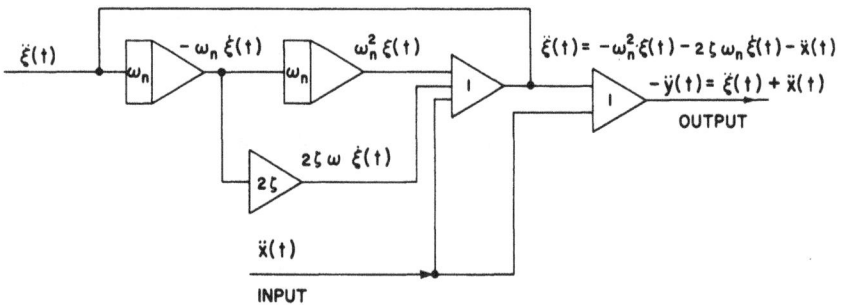

Fig. 5.3.   Rearranged circuit for the analog of a simple mechanical oscillator

Two integrators and three summing amplifiers thus perform the same operation as four integrators and four summing amplifiers. The negative of the mass acceleration is obtained by the circuit of Fig. 5.3.

Usually, the correct polarity can be restored by the appropriate connection of the plotting device. If not, an additional unity gain summing amplifier is required to resolve the polarity.

If certain restrictions on the minimum damping are permissible, the circuit can be further reduced to that shown in Fig. 5.4 [24].

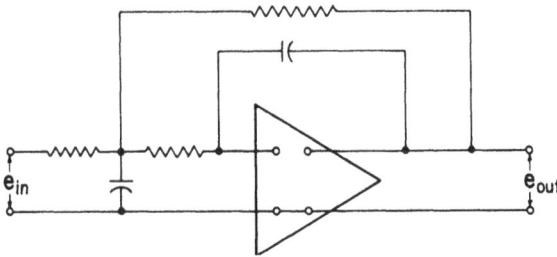

Fig. 5.4. Single-amplifier analog circuit of a damped mechanical oscillator.

Passive Analog Computers

Passive analog computers are primarily intended for the solution of problems involving the response of physical systems because their operation is based on the similarity of the equations describing the behavior of electrical elements and physical elements. A passive electrical circuit is formed that simulates the physical system. The differential equations describing the performance of the electrical system must be proportional to the differential equations describing the performance of the physical system. The primary computing elements in passive analog computers are resistors, capacitors, and inductors (transformers, operational amplifiers, and function generators are also required for many problems).

Passive analog computers are particularly well suited to the solution of problems involving the response of discrete and distributed structures. This results from the similarity between Newton's laws describing the behavior of the structure and Kirchoff's laws describing the behavior of the circuit. There are two different electrical analogies, and they are based on the two different laws of Kirchoff. These are

1. The amount of current flowing out of a point (node) in a circuit is equal to the amount of current flowing into the point.

2. The difference of electrical potential between any two points in a circuit is independent of the path used to measure it. (This is more commonly stated as the sum of voltage drops in a closed loop is equal to the sum of the voltage rises in the loop.)

Kirchoff's first law is the basis for the nodal analogy. The sum of all the currents injected into a node is equal to the sum of all the nodal admittances multiplied by the potential difference of the node. The currents in single, passive-element nodes are as follows.

Resistor	Time Domain	Frequency Domain

$$i_R(t) = \frac{e_1(t) - e_2(t)}{R} \qquad I_R(f) = \frac{E_1(f) - E_2(f)}{R}$$

Capacitor

$$i_C(t) = C\frac{d[e_1(t) - e_2(t)]}{dt} \qquad I_C(f) = j2\pi f C\,[E_1(f) - E_2(f)]$$

Inductor

$$i_L(t) = \int_{-\infty}^{t} \frac{[e_1(\tau) - e_2(\tau)]d\tau}{L} \qquad I_L(f) = \frac{E_1(f) - E_2(f)}{j2\pi f L}$$

The loop analogy is based on Kirchoff's second law. The sum of the applied voltages in a loop is equal to the sum of the loop impedance multiplied by the net loop current through each impedance. The voltage across a single passive element is found as follows.

Resistor	Time Domain	Frequency Domain

$$e_R(t) = [i_1(t) - i_2(t)]R \qquad E_R(f) = [I_1(f) - I_2(f)]R$$

Capacitor

$$e_C(t) = \int_{-\infty}^{t} \frac{[i_1(\tau) - i_2(\tau)]d\tau}{C} \qquad E_C(f) = \frac{I_1(f) - I_2(f)}{j2\pi f C}$$

Inductor

$$e_L(t) = L\frac{d[i_1(t) - i_2(t)]}{dt} \qquad E_L(f) = j2\pi f L[I_1(f) - I_2(f)]$$

To illustrate the application of the passive electrical analogies to the solution of practical problems, again consider the base-acceleration-excited, simple mechanical oscillator. The equation of motion (see Fig. 5.1) is

$$m\ddot{y} + c(\dot{y} - \dot{x}) + k(y - x) = 0. \qquad (5.3)$$

It will again be convenient to express this equation in terms of the motion of the mass relative to the base $[\xi(t) = y(t) - x(t)]$,

$$m\ddot{\xi} + c\dot{\xi} + k\xi = -m\ddot{x}. \qquad (5.4)$$

Or, in terms of Laplace transforms,

$$s^2 mZ(s) + scZ(s) + kZ(s) = s^2 mX(s). \qquad (5.5)$$

Nodal Analogy

First, the nodal electrical analogy will be considered. If currents in the electrical analog are made proportional to mechanical forces, the appropriate electrical equation is

$$sCE_0(s) + \frac{E_0(s)}{R} + \frac{E_0(s)}{sL} = I_0(s). \qquad (5.6)$$

This equation yields the circuit of Fig. 5.5.

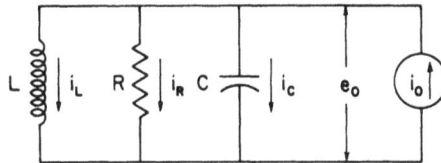

Fig. 5.5.   Nodal analogy of a simple mechanical oscillator.

The comparison of the mechanical and electrical systems is helped if Eq. (5.5) is rewritten in terms of the relative velocity:

$v(t) = \dot{y}(t) - \dot{x}(t) =$ the relative velocity of the mass to the base

$V(s) = sZ(s) =$ the Laplace transform of the relative velocity

$$smV(s) + cV(s) + \frac{kV(s)}{s} = s^2 mX(s). \qquad (5.7)$$

A comparison of Eqs. (5.6) and (5.7) shows the following relation:

$V(s)$, or $v(t)$ (relative velocity) $\sim E_0(s)$, or $e_0(t)$ (voltage)

$smV(s)$ (inertial force) $\sim sCE_0(s)$ (current through the capacitor)

$cV(s)$ (damping force) $\sim \dfrac{E_0(s)}{R}$ (current through the resistor)

$\dfrac{kV(s)}{s}$ (spring force) $\sim \dfrac{E_0(s)}{sL}$ (current through the inductor)

$-sm^2X(s)$ (input "force") $\sim I_0(s)$ (input current).

From the above comparison these additional relations can be derived;

m (mass) $\sim C$ (capacitance)

c (damping constant) $\sim \dfrac{1}{R}$ (reciprocal of resistance $=$ conductance)

k (spring constant) $\sim \dfrac{1}{L}$ (reciprocal of inductance).

To summarize, the nodal analog of the base-acceleration-excited, simple mechanical oscillator is formed by selecting a capacitor proportional to the mass of the oscillator, a resistor that is proportional to the reciprocal of the damping constant of the oscillator, and an inductor that is proportional to the reciprocal of the spring constant of the oscillator. These three electrical elements are connected in parallel as shown in Fig. 5.5. An input current is applied that is proportional to the base acceleration scaled by a factor of minus one times the mass of the oscillator.

The desired output was stated as the acceleration of the mass. This acceleration is proportional to the difference between the current in the capacitor and the input current, scaled by a factor equal to the mass of the oscillator. As shown below, it also equals the product of $-1/m$ and the sum of the currents in the resistor and the inductor;

$$\ddot{y}(t) = \ddot{\xi}(t) + \ddot{x}(t)$$
$$m\ddot{y}(t) = m\ddot{\xi}(t) + m\ddot{x}(t) \sim i_c(t) - i_0(t) = -i_R(t) - i_L(t). \tag{5.8}$$

Because current is proportional to force and voltage is proportional to velocity, the nodal analogy is sometimes known as the force-current, velocity-voltage analogy.

Loop Analogy

Next, consider the loop analogy. In this case, the appropriate electrical equation is

$$E_0(s) = sLI_0(s) + RI_0(s) + \frac{I_0(s)}{sC}, \tag{5.9}$$

which describes the circuit shown in Fig. 5.6.

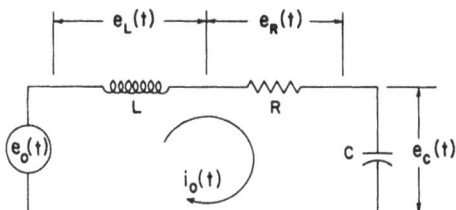

Fig. 5.6.   Loop analogy of a simple
mechanical oscillator.

A comparison of the equations describing the mechanical system, Eq. (5.7), and the electrical system, Eq. (5.9), reveals the following relations:

$smV(s)$ (inertial force) $\sim sLI(s)$ (voltage across the inductor)

$cV(s)$ (damping force) $\sim RI(s)$ (voltage across the resistor)

$\dfrac{kV(s)}{s}$ (spring force) $\sim \dfrac{I(s)}{sC}$ (voltage across the capacitor)

$V(s)$ (relative velocity) $\sim I(s)$ (loop current)

m (mass) $\sim L$ (inductance)

c (damping constant) $\sim R$ (resistance)

k (spring constant) $\sim \dfrac{1}{C}$ (reciprocal of capacitance)

$-ms^2X(s)$ (input "force") $\sim E_0(s)$ (input voltage).

To summarize, the loop analogy of the base-acceleration-excited, simple mechanical oscillator is formed by selecting an inductor proportional to the mass of the oscillator, a resistor proportional to the damping coefficient, and a capacitor that is proportional to the reciprocal of the spring constant. These electrical elements are then connected in series as shown in Fig. 5.6. An input voltage is applied that is proportional to the base acceleration scaled by a factor equal to the mass of the mechanical oscillator.

Again, the desired output is the absolute acceleration of the mass. This acceleration is proportional to the sum of the input voltage and the voltage across the inductor, scaled by a factor proportional to the mass.

As shown below, it is also equal to the product of $-1/m$ and the sum of the voltage across the resistor and capacitor;

$$\ddot{y} = \ddot{\xi} + \ddot{x}$$

$$m\ddot{y} = m\ddot{\xi} + m\ddot{x}$$

$$e_{\text{out}}(t) = e_L(t) + e_0(t) = -e_R(t) - e_C(t). \tag{5.10}$$

Because the voltages are proportional to force and the loop current is proportional to the velocity, this loop analogy is sometimes known as the force-voltage, velocity-current analogy.

The choice of a nodal or a loop analogy depends on the specific application of interest, as each analogy has its specific advantages and disadvantages. The nodal analogy has the advantage of being topologically similar to the mechanical system. In many cases, the analog of the mechanical system can be created just by replacing masses by capacitors, etc. On the other hand, the loop analogy has an advantage in that the parameters of interest are usually voltages, whereas they are usually described by currents in the nodal analogy. This is an advantage because it is easier to measure voltages without disturbing the circuit than it is to measure currents.

5.2 Computation of Fourier Spectra

Analog instruments that are used to compute Fourier transforms are divided into a number of different categories. One division is that of *parallel* vs *swept* analysis. Parallel analyzers compute all the spectral values of interest simultaneously. They perform a true real-time analysis. Swept analyzers compute only one spectral value at a time. The swept analyzers are divided into groups depending on the manner in which the frequency of the spectral computation is changed. In one type, this frequency is stepped and, in the other type, the frequency is continuously varied. (In both types, the frequency must be changed slowly to prevent error.) The performance of a swept type of Fourier analysis requires that the data to be analyzed must be periodic. Data that are not naturally periodic (such as single pulse data) are forced to appear periodic to the analyzer. This artificial periodicity is usually created by recording the data on a continuous loop of magnetic tape. This forces the data to be periodic with a fundamental frequency equal to the reciprocal of the duration of a tape loop. The parallel analyzer has an obvious speed advantage and price disadvantage (for equal-quality analyses). Analyzers may also be divided into those which directly implement the mathematical equations describing the Fourier transform and those which employ bandpass filters.

Mathematical Equation Implementation

The direct mathematical equation implemented is the trigonometric version of the Fourier transform as related in Eq. (5.11),

$$X(f) = \int_{-\infty}^{\infty} x(t) \cos 2\pi ft\, dt - j \int_{-\infty}^{\infty} x(t) \sin 2\pi ft\, dt. \tag{5.11}$$

The block diagram of an analyzer based on the above equation is shown in Fig. 5.7. This circuit computes the real and imaginary parts of the Fourier transform at one frequency. To obtain the entire spectrum either the circuit must be duplicated (parallel analysis) or the frequency of the oscillator must be changed. Typical plots from this type of analyzer are shown in Fig. 5.8.

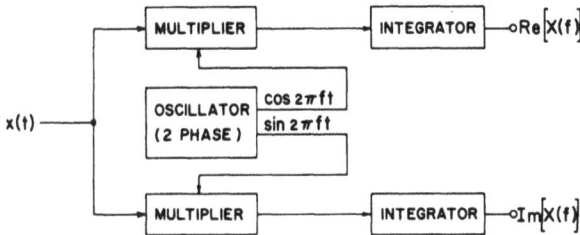

Fig. 5.7. Block diagram of an analog Fourier analyzer (direct mathematical model).

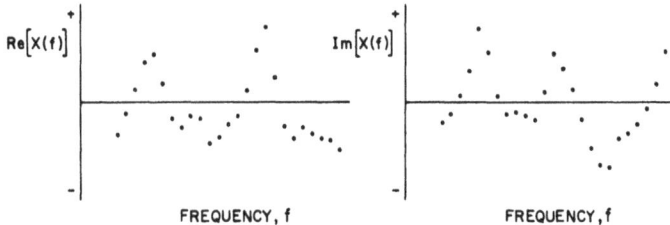

Fig. 5.8. Typical plots of the real and imaginary portions of a Fourier spectrum.

Sometimes it is more convenient to express the Fourier transform in terms of its modulus and phase factor. The block diagram of the circuitry used to obtain these values from the real and imaginary values is shown in Fig. 5.9, and the quantities are described mathematically in Eq. (5.12);

$$|X(f)| = \sqrt{Re^2[X(f)] + Im^2[X(f)]} \tag{5.12a}$$

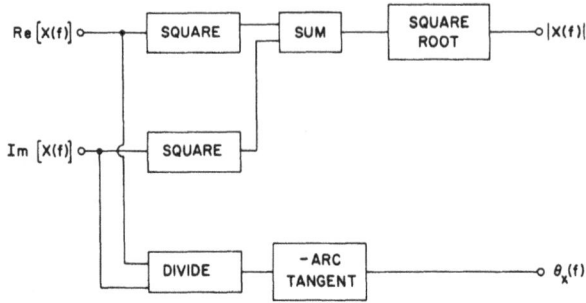

Fig. 5.9. Basic block diagram for converting real and imaginary forms of Fourier transforms to modulus and phase spectra.

$$\theta_x(f) = -\tan^{-1}\left\{\frac{Im[X(f)]}{Re[X(f)]}\right\}. \tag{5.12b}$$

These values can be plotted either in cartesian form as shown in Fig. 5.10 or in polar form as shown in Fig. 5.11.

This type of analyzer works quite well; however, as in any practical analysis, there are certain sources of error that must be kept in mind. The first requirement is that the oscillators must be keyed to start at exactly the same time as the transient. Any phase shift between the start of the transient and the 0° value of the sinusoids will be reflected directly in the phase factor. The modulus will not be affected by this error source. However, if real and imaginary values are plotted instead of modulus and phase values, both will be in error. In fact, there will be cross-coupling between the real and imaginary portions of the true Fourier transform.

To further illustrate this error, the Fourier transform of a cosine wave will be computed over N integral cycles. Assume that the sinusoidal generators are free-running and that the oscillators are shifted in phase by θ_0 degrees when the cosine wave is started. The relative timing is shown in Fig. 5.12.

The equations required to find the Fourier transform are as follows:

$$X(f_0) = \int_0^{N/f_0} \cos(2\pi f_0 t)\cos(2\pi f_0 t + \theta_0)\,dt$$

$$-j\int_0^{N/f_0}\cos(2\pi f_0 t)\sin(2\pi f_0 t + \theta_0)\,dt$$

$$=\frac{\cos(-\theta_0)}{2}\frac{N}{f_0} + j\frac{\sin(-\theta_0)}{2}\frac{N}{f_0} \tag{5.13a}$$

$$M_x(f_0) = \sqrt{Re^2[X(f_0)] + Im^2[X(f_0)]} = \frac{N}{2f_0} \tag{5.13b}$$

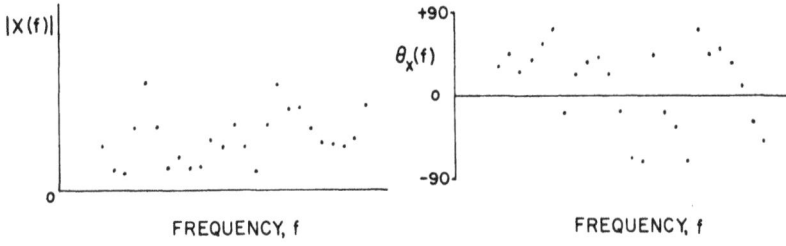

Fig. 5.10. Typical plot of the modulus and phase factors of Fourier spectra.

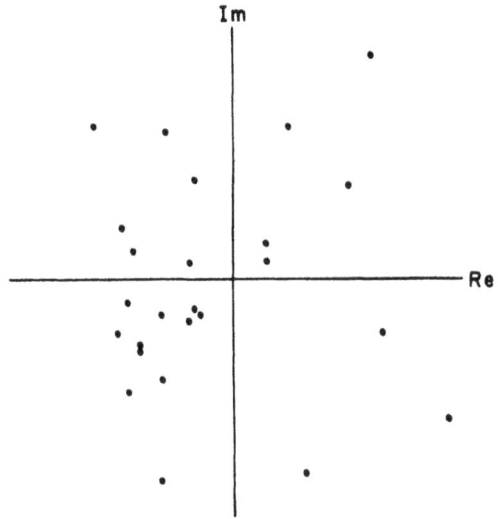

Fig. 5.11. Typical polar plot of a
Fourier spectrum.

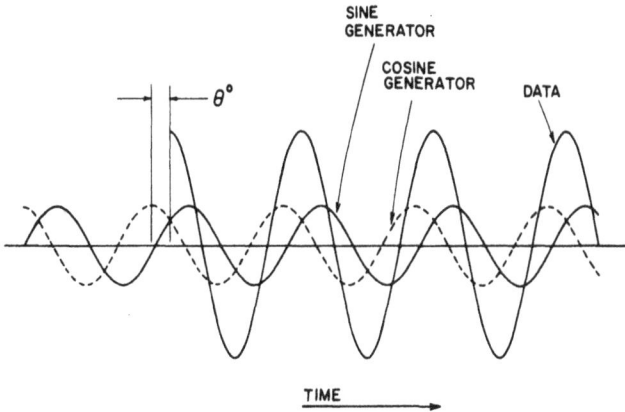

Fig. 5.12. Timing diagram.

$$\theta_x(f_0) = -\tan^{-1}\left[\frac{Im[X(f_0)]}{Re[X(f_0)]}\right] = \theta_0. \qquad (5.13c)$$

When $\theta_0 = 0$, the correct value, $X(f) = \dfrac{N}{2f_0}$, is obtained.

The cross-coupling in the real and imaginary values, when this phase error exists, can be illustrated as follows:

$$Re'[X(f)] = M \cos \theta' = M \cos (\theta + \alpha) \qquad (5.14a)$$

$$Im'[X(f)] = M \sin \theta' = M \sin (\theta + \alpha), \qquad (5.14b)$$

where

$Re'[X(f)] =$ the indicated real portion of the Fourier transform

$Im'[X(f)] =$ the indicated imaginary portion of the Fourier transform

$\alpha =$ the phase error

$$Re'[X(f)] = M (\cos \theta \cos \alpha + \cos \theta \sin \alpha) \qquad (5.15a)$$
$$= Re[X(f)] \cos \alpha - Im[X(f)] \sin \alpha$$

$$Im'[X(f)] = M (\sin \theta \cos \alpha + \cos \theta \sin \alpha) \qquad (5.15b)$$
$$= Im[X(f)] \cos \alpha + Re[X(f)] \sin \alpha,$$

where

$Re[X(f)] =$ the true real portion of the Fourier transform

$Im[X(f)] =$ the true imaginary portion of the Fourier transform.

This source of error prevents the use of continuous-frequency sweep analyses with this type of analyzer. The circuits must be duplicated or the frequency of the oscillators must be stepped — exercising care that zero phase angle corresponds to the start of the transient at each frequency.

A second source of error with this type of analyzer is harmonic distortion in the electronic circuits. The largest contributors to this type of error are generally the multiplier circuits, although care must be exercised to insure that the oscillators do not have significant energy at frequencies other than the fundamental one. The ability of the multiplying circuitry to perform a true multiplication will generally be the limiting element.

Estimating the magnitude of this type of error is quite difficult, as the error depends not only on imperfections in the hardware but also on the shape of the Fourier transform. As an example, assume that the oscillators have some higher order, odd harmonic components. Then the actual Fourier transform computed by the instrument is as follows:

$$X^m(f_0) = \int_0^T x(t) \, (\cos 2\pi f_0 t + A_1 \cos 6\pi f_0 t + A_2 \cos 10\pi f_0 t + \ldots) \, dt$$

$$-j \int_0^T x(t) \, (\sin 2\pi f_0 t + A_1 \sin 6\pi f_0 t + A_2 \sin 10\pi f_0 t + \ldots) \, dt$$

$$= X(f_0) + A_1 X(3f_0) + A_2 X(5f_0) + \ldots, \tag{5.16}$$

where

$X^m(f_0)$ = the value of the Fourier transform displayed at the output of the analyzer

A_n = the relative magnitude of the nth harmonic of the oscillator to the fundamental.

An examination of Eq. (5.16) shows that the displayed value of the Fourier transform contains the true value, plus scaled values of the Fourier transform at odd multiples of f_0. When the value of the true Fourier transform is much larger at a distortion frequency (for example at $3f_0$ in the above example) than the value of the true Fourier transform at the analysis frequency (f_0 in the above example), the displayed value can be grossly in error for what would appear to be small values of harmonic distortion. If, in the above example, the value of the true Fourier transform at $3f_0$ was 100 times larger than the value of the true Fourier transform at f_0, and A_1 had a value of 0.01 (1-percent distortion at this frequency) then the indicated value of the Fourier transform at f_0 would be twice the true value.

Bandpass Filter Implementation

The second type of analog Fourier analyzer utilizes narrow bandpass filters followed by detectors as shown in Fig. 5.13.

Fig. 5.13.   Block diagram of an analog Fourier analyzer (bandpass filter model).

For periodic data, the principle on which they operate is fairly easy to visualize. Only one spectral component is allowed to pass through the filter at a time. All other spectral components are removed from the output of the filter. In Fig. 5.14 the dotted line symbolizes an ideal bandpass filter centered at frequency f_0. The output of the filter will be a sine wave with an rms value of M_0 (assuming unity gain in the filter passband).

Since the output signal is a sine wave, many types of detectors can be employed to detect its rms value. True rms detectors, peak detectors, or mean absolute value detectors (as shown in Fig. 5.13) are commonly used. Any of these detectors is perfectly acceptable as long as the appropriate calibration factor is accounted for in the final results. The

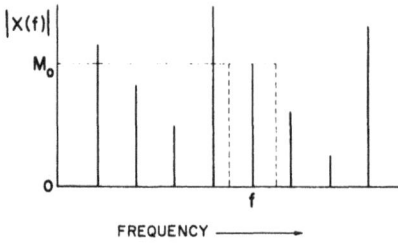

Fig. 5.14.   Spectrum of a periodic function.

calibration factor is the ratio of the sine wave characteristic measured to the rms value of a sine wave. For a peak value measurement the calibration factor is 0.707. ($E_{rms}/E_{peak} = 0.707$ for a sine wave.) For a mean absolute value measurement (ordinary AC voltmeter) the calibration factor is 1.11. ($E_{rms}/E_{mean} = 1.11$ for a sine wave.)

With an actual filter instead of the idealized one previously assumed, the output will have some contributions from all spectral components of the input, as shown in Fig. 5.15.

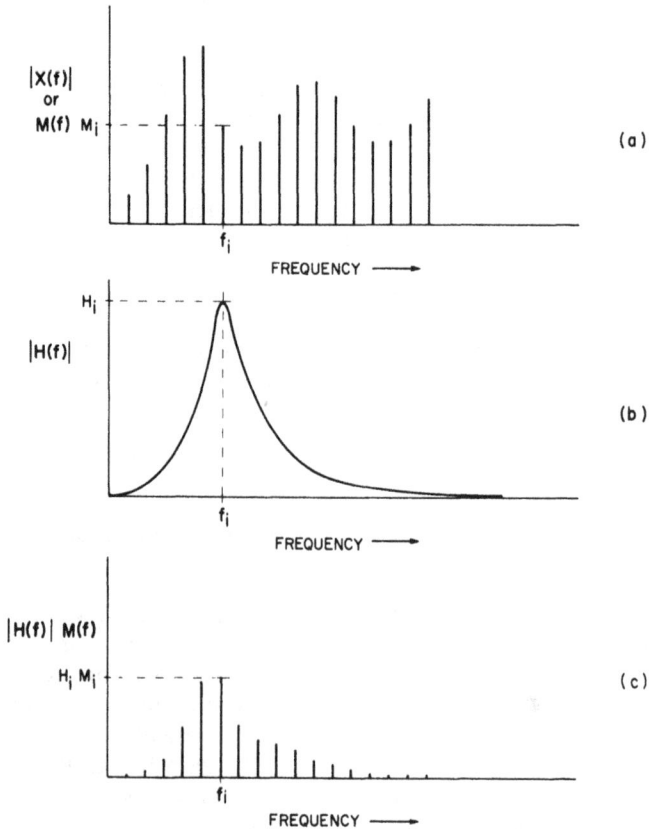

Fig. 5.15.   The effect of a nonideal filter on Fourier analysis; (a) modulus of the input, (b) filter gain factor, (c) spectral contributions (moduli) to the filter output.

Thus it is important that the filter have a steep rolloff rate, so that it strongly attenuates all but one spectral component. The total input of a periodic signal can be represented as follows:

$$x(t) = \sum_{i=0}^{n} M_i \cos (2\pi f_i + \theta_i). \tag{5.17}$$

At each spectral frequency, the amplitude of the cosine wave is multiplied by the gain of the filter at that frequency, and the phase of the cosine wave is shifted by the amount of phase shift through the filter at that same frequency. Thus the output of the filter is

$$y(t) = \sum_{i=0}^{n} H_i M_i \cos (2\pi f_i + \theta_i + \phi_i) \tag{5.18}$$

where
$\quad H_i =$ the gain of the filter at the ith frequency
$\quad \phi_i =$ the phase shift through the filter at the ith frequency.

Calculating the rms value of the filter output yields

$$\sqrt{\overline{y^2}} = \sqrt{\sum_{i=0}^{n} H_i^2 M_i^2}. \tag{5.19}$$

The indicated value of the modulus of the Fourier component at f_0 is the mean square value normalized by the gain of the filter at f_0;

$$M_{f_0}^m = \frac{\sqrt{\overline{y^2}}}{H_{f_0}} = \sqrt{\sum_{i=0}^{n} \left(\frac{H_i}{H_{f_0}}\right)^2 M_i^2}. \tag{5.20}$$

The measured value will differ from the true value. The difference, or error term is defined in the following equation:

$$e = \frac{M_{f_0}^m - M_{f_0}}{M_{f_0}} = \frac{\left[\sqrt{\sum_{i=0}^{n} \left(\frac{H_i}{H_{f_0}}\right)^2 M_i^2}\right] - M_{f_0}}{M_{f_0}}$$

$$= \left[\sqrt{\sum_{i=0}^{n} \left(\frac{H_i}{H_{f_0}}\right)^2 \left(\frac{M_i}{M_{f_0}}\right)^2}\right] - 1 = \left[\sqrt{1 + \sum_{i \neq f_0} \left(\frac{H_i}{H_{f_0}}\right)^2 \left(\frac{M_i}{M_{f_0}}\right)^2}\right] - 1.$$

Using the binomial expansion,

$$e \approx 1 + \frac{1}{2} \sum_{n \neq f_0} \left(\frac{H_i}{H_{f_0}}\right)^2 \left(\frac{M_i}{M_{f_0}}\right)^2 - 1 \approx \frac{1}{2} \sum_{n \neq f_0} \left(\frac{H_i}{H_{f_0}}\right)^2 \left(\frac{M_i}{M_{f_0}}\right)^2. \tag{5.21}$$

This expression is generally quite difficult to evaluate, since the true spectrum must be known. However, Eq. (5.21) does demonstrate the need for the bandpass filter in the analyzer to have very low gain for all spectral components except the one at the center frequency of the filter.

Periodic Data Considerations

Frequently, complicated data are artificially made periodic, such as by recording the transient on magnetic tape, forming the tape into a continuous loop, and then recirculating the loop many times for single-filter, scanned analysis. In this case there is likely to be a number of spectral components within the passband of the filter used for analysis. The output indicated by the analyzer at the center frequency will be the sum of all the spectral components over the passband. In this manner a discrete measurement approximating the true continuous spectrum of the transient is obtained. (By discrete measurement, it is meant that a single point is obtained for each center frequency setting of the filter.)

As an example, consider the terminal-peak sawtooth pulse shown in Fig. 5.16a. The amplitude spectrum of this pulse is shown in Fig. 5.16b. If the data are made periodic by connecting the end of the transient to its beginning, the time history will appear as shown in Fig. 5.16c. In this case, the continuous spectrum is estimated by discrete points separated in frequency by the reciprocal of the duration of the transient. For example, if the sawtooth pulse in the preceding figure has a duration of 10 msec, spectral components will be 100 Hz apart.

Frequently, it is impossible to make a single pulse repetitive by splicing the end of the transient to its beginning. Practical considerations, such as the minimum loop length that the tape recorder can handle, or the desire to blank out tape splice noise are examples of such factors. In this case, the time history resembles the example shown in part e of Fig. 5.16. The duration of the transient is still T_0, but it is now repeated once every T_1 seconds. When the transient is made periodic in this manner, the spectrum consists of discrete values spaced $(1/T_1)$ Hz apart. For example, the same 10-msec transient might be recorded on a loop of magnetic tape that recirculates completely once every 2 sec. Then the spectral spacing would be 1/2 Hz. Note that the magnitude of any spectral component is equal to the value of the continuous spectrum of the nonrepeated transient evaluated at that particular frequency.

If the filter bandwidth used is greater than $1/T$, as is usually the case, then more than one spectral component will occur within its passband. The indicated output of the analyzer will then be just the average value of these spectral components as weighted by the filter's frequency response function. An error between the indicated value and the true value at the center frequency of the filter is possible, even when assuming an idealized rectangular bandpass filter shape. Consider the three spectral values in Fig. 5.17 as being a portion of an overall spectrum bandpassed through an idealized filter created at frequency

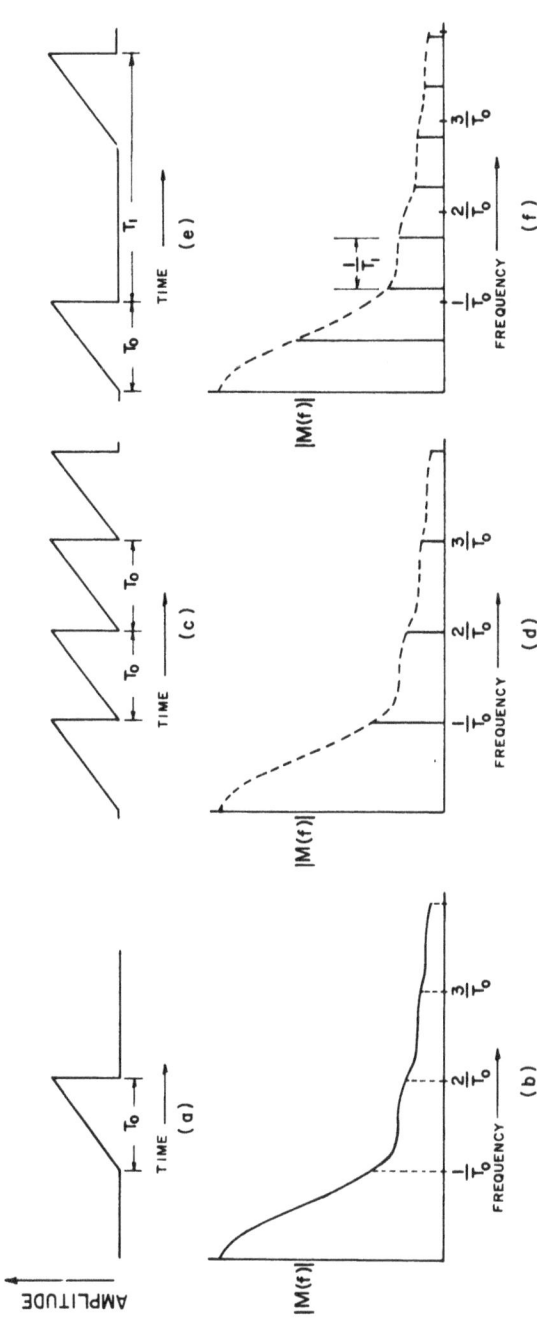

Fig. 5.16. Single and repeated triangular pulses and their spectra: (a) terminal-peak sawtooth pulse, (b) spectrum of (a), (c) (a) repeated continuously, (d) spectrum of (c), (e) spacing between pulses, (f) spectrum of (e).

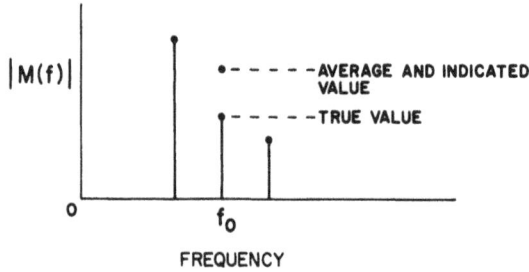

Fig. 5.17. Error in spectral measurement due to
finite filter bandwidth.

f_0. The true modulus at f_0 is lower than the average value of the three
spectral components.

The magnitude of this error can be estimated (as shown in Ref. 25), as

$$E(f_0) \approx \frac{B^2}{24} |\ddot{M}(f_0)|, \qquad (5.22)$$

where

$B =$ the bandwidth of the filter

$|\ddot{M}(f_0)| =$ the second derivative of the modulus of the
Fourier spectrum with respect to frequency
evaluated at f_0.

It should be noted that there will be no error when the spectrum varies
linearly over the filter bandpass. Also, since the error formula results
from dropping higher order terms in a Taylor series expansion, its
value in estimating the magnitude of the error diminishes rapidly when
the spectrum is not monotonic inside the filter passband.

Scan Rate Considerations

There is another potential source of error in analog Fourier analyzers
employing a single continuously scanned bandpass filter. This error
occurs when the single filter is scanned too rapidly. Rapid scanning
causes the effective frequency response function of the filter to differ
drastically from its static (unscanned) state.

The effect of rapid scanning of a filter is shown in Fig. 5.18.

The most prominent distortions of the scanned filter characteristics
are a decrease in the peak response and the shifting of the peak
response frequency in the scanning direction (increasing frequency
scan in the figure). Kharkevich [26] presents the following formula for
estimating the magnitude of the scanned filter errors (a second order
filter is assumed):

Peak Response

$$e_p = \frac{-4\lambda^2}{\pi^2 B^4}, \qquad (5.23)$$

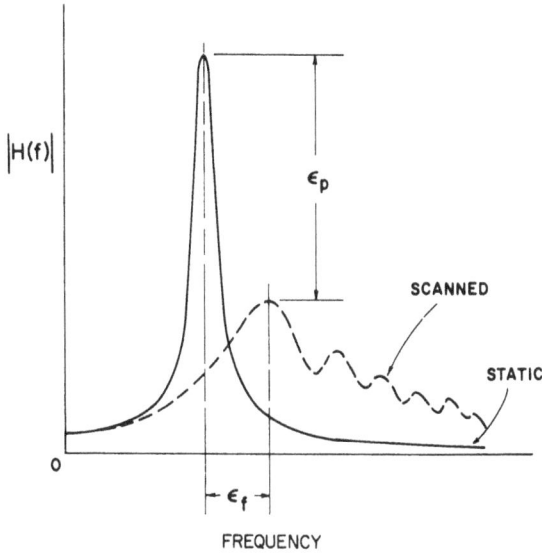

Fig. 5.18.   Gain factors of a filter
(static and scanned).

where

$$e_p = \text{the fractional error in the peak response}$$

$$\lambda = \text{the scan rate in Hz/sec} = \frac{\Delta f}{T}$$

$$B = \text{the bandwidth of the filter.}$$

Frequency Shift

$$e_f = \frac{4\lambda}{\pi B^2}, \tag{5.24}$$

where

$$e_f = \text{the frequency shift error in Hz.}$$

Bandpass Filtering Techniques

Three different techniques are used in this type of analyzer to perform the bandpass filtering technique. The first technique is *direct filtering*. The analyzer employs a separate filter for each frequency to be analyzed. The filter has a center frequency equal to the desired analysis frequency and has the desired bandwidth. Figure 5.19 a and b illustrate the spectrum and direct filtering. The dotted line in part (b) represents the filter frequency response. The second filtering technique is *heterodyne filtering*. In this technique, the data amplitude modulates a high frequency sinusoidal carrier signal. Since suppressed carrier modulation is normally employed, the data can be considered as multiplying the carrier signal. The resultant signal $e_0(t)$ is given by the following equation:

$$e_0(t) = (\text{data}) \times (\text{carrier})$$

$$= (|M_0| + |M| \cos (2\pi f_1 t + \theta_1) + \ldots$$

$$+ |M_n| \cos (2\pi n f_1 t + \theta_n))(A \cos 2\pi f_c t). \quad (5.25a)$$

This can be expanded as follows:

$$e_0(t) = A \left\{ |M_0| \cos 2\pi f_c t + \frac{|M_1|}{2} \cos [2\pi (f_c + f_1)t + \theta_1] + \ldots \right.$$

$$+ \frac{|M_n|}{2} \cos [2\pi (f_c + f_n)t + \theta_n] + \frac{|M_1|}{2} \cos [2\pi (f_c - f_1)t - \theta_1] + \ldots$$

$$\left. + \frac{|M_n|}{2} \cos [2\pi (f_c - f_n)t - \theta_n] \right\}. \quad (5.25b)$$

This resultant signal has the original spectrum folded about the carrier frequency. To perform the filtering, a filter with a high center frequency is used. Either the sum or difference frequency can be filtered. Part (c) of Fig. 5.19 shows the filter operating on the difference frequency spectrum.

There are two reasons for using the heterodyne filtering method. First, very sharp bandpass filters can be realized at these frequencies by using quartz or magnetostrictive filters. Second, a single filter can effectively be stepped or scanned through the data simply by changing the frequency of the carrier signal, and the characteristics of the filter need not be changed. When the frequency of the carrier is changed, a new portion of the difference frequency spectrum falls within the pass-band of the filter in part (c). In actuality, the data frequency spectrum is scanned past the filter rather than the filter being scanned past the data.

The third technique is called *homodyne filtering*. This technique is quite similar to heterodyne filtering. However, in this technique, the data signal is multiplied by a carrier signal having a frequency *equal* to the frequency at which the spectrum is to be analyzed. The multiplication again results in sum and difference frequency spectra. However, the center frequency of the analysis filter now corresponds to a difference frequency of zero. Thus, the filter actually used is a low-pass filter; and since the negative frequency difference components have their energy folded around zero into the positive frequency region, the actual filter need only have half of the desired bandwidth. Part (d) of Fig. 5.19 shows a low-pass filter of $B/2$ Hz bandwidth effectively performing a bandpass filtering of the data spectrum at f_0 and with bandpass B Hz.

As with the heterodyne filter, the data can be moved past a single filter by changing the frequency of the carrier. Use of the low-pass

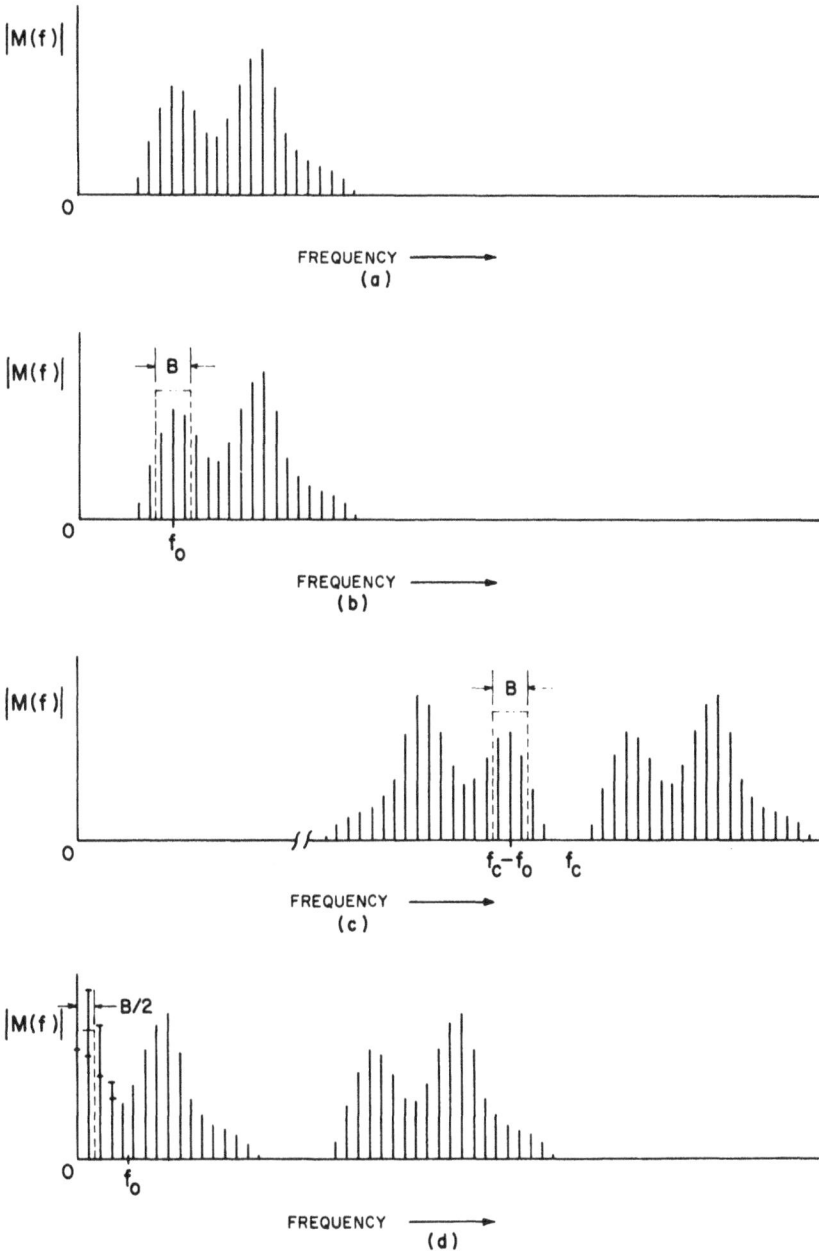

Fig. 5.19. Filtering techniques; (a) basic spectrum, (b) direct filtering, (c) heterodyne filtering, (d) homodyne filtering.

filter removes a source of error, frequency drift, from the filter but not from the carrier signal generator. There is one other factor that must be corrected for in homodyne filtering. That is the relative phasing between the data signal and the carrier signal. Examining a single spectral component in the product signal, it is found that the resultant difference frequency is a dc value whose amplitude is proportional to the amplitude of the spectral component multiplied by the cosine of the phase difference between the spectral component and the carrier frequency;

$$e_0(t) = [|M(f_0)| \cos (2\pi f_0 t + \theta_0)] \cos (2\pi f_0 t + \theta_c)$$

$$= \frac{|M(f_0)|}{2} [\cos (\theta_0 - \theta_c)]. \tag{5.26}$$

This value can range anywhere from zero to the correct value of $|M(f_0)|/2$. (Only the absolute value of the cosine is of importance, as polarity is lost in the detector.)

This problem could be approached by synchronizing the carrier oscillator; however, this would preclude the use of continuous scanning. Instead, the data are multiplied by both the sine and cosine of the analysis frequency as shown in Fig. 5.20. Each multiplier signal is passed through a separate low-pass filter. For a single spectral component, this yields the difference frequency signals of $[|M(f_0)|/2]$ $\cos (\theta_0 - \theta_c)$ and $[|M(f_0)|/2] \sin (\theta_0 - \theta_c)$ in the upper and lower channels, respectively.

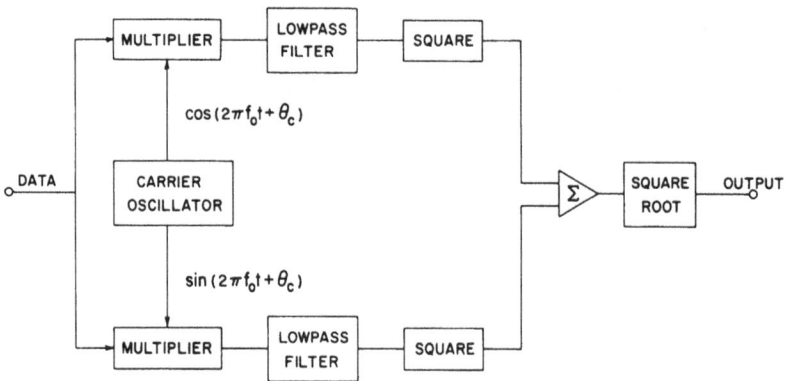

Fig. 5.20.   Practical homodyne filtering.

The output is computed by squaring each filter output, adding the squared values, and then taking the square root of the output;

$$\text{Output} = \sqrt{\left[\frac{|M(f_0)|}{2} \cos (\theta_0 - \theta_c)\right]^2 + \left[\frac{|M(f_0)|}{2} \sin (\theta_0 - \theta_c)\right]^2}$$

$$= \frac{|M(f_0)|}{2} \sqrt{\cos^2(\theta_0 - \theta_c) + \sin^2(\theta_0 - \theta_c)}$$

$$= \frac{|M(f_0)|}{2}. \tag{5.27}$$

Thus, the output is equal to the correct value of the spectrum at frequency f_0. (The factor of two is accounted for in the calibration.)

When the filter passband enclosed only one spectral component, it was noted that a true rms, mean absolute value, or peak detector could be used, since the output waveform from the filter is a single sinusoid, and the mean absolute value and peak value of a sinusoid are readily converted to a true rms reading by applying a simple calibration constant. When there are two or more spectral components enclosed by the passband of the analyzer, a true rms detector must be employed. Peak and mean absolute value detectors will give erroneous readings of the rms value.

To illustrate this fact, consider the simple example of a sine wave and its second harmonic, with no initial phase shift. The rms value is found as follows:

$$\sqrt{\Psi_x^2} = \frac{1}{T} \left[\int_0^T (\sin 2\pi ft + k \sin 4\pi ft)^2 dt \right]^{1/2}. \tag{5.28a}$$

Let $T = f/4$, since the signal is periodic every quarter of a cycle of the fundamental frequency;

$$\sqrt{\Psi_x^2} = \left[4f \int_0^{f/4} (\sin^2 2\pi ft + 2k \sin 2\pi ft \sin 4\pi ft + k^2 \sin^2 4\pi ft) dt \right]^{1/2}$$

$$= \frac{\sqrt{1 + k^2}}{\sqrt{2}}. \tag{5.28b}$$

The mean absolute value is found as follows:

$$|\bar{X}| = \frac{1}{T} \int_0^T |\sin 2\pi ft + k \sin 4\pi ft| \, dt. \tag{5.29a}$$

Again the same upper limit can be used, and since both signals are positive out to $T = f/4$, the absolute value can be dropped;

$$|\bar{X}| = 4f \int_0^{f/4} (\sin 2\pi ft + k \sin 4\pi ft) \, dt = \frac{2}{\pi}(1 + k). \tag{5.29b}$$

If the mean absolute value could be used in place of the true rms value, then the ratio of the true rms to it would be a fixed calibration constant;

$$\frac{\sqrt{\Psi_x^2}}{|\bar{X}|} = \frac{\frac{1}{\sqrt{2}}\sqrt{(1+k^2)}}{\frac{2}{\pi}(1+k)} = \text{constant}. \tag{5.30}$$

Since this ratio is a function of k, the mean absolute value detector cannot be calibrated in terms of the true rms value. (The ratio is also a function of any phase angle between the spectral components.)

RC Averaging Techniques

Not all analog Fourier analyzers employing bandpass filters use true integrators. In fact, the majority use RC averaging in place of true integration. This is legitimate, as long as certain precautions are observed in the use of the RC averager, because the integrator is being used to compute a time-averaged value. This operation is essentially that of low-pass filtering. The dc voltage on the output of the detector is the quantity of interest. Figure 5.21a shows the block diagram of a true integrator and an RC averager. Part (b) shows the gain factor of these two circuits. The RC averager is the simplest low-pass filter available. It also gives a fairly good approximation to the true integrator, since the envelopes of both gain factors are proportional to $1/f$ at high frequencies.

The precaution to be observed in using an RC averager to replace the true integrator can be studied by examining Fig. 5.21c and d. Part (c) compares the weighting functions of the two circuits. By use of these weighting functions and the convolution integral, the response of each circuit can be computed. When an input is applied, a step in the mean square value occurs at the output of the filter (and also out of the squaring circuit). When this step is applied to the integrator, its output responds by building up linearly with time. Thus, the input can be determined at any time t_0 simply by multiplying the output of the integrator by (T/t_0).

With the RC averager, the output gets closer to the product of $1/RC$ and the input as the averaging time increases. The idea is to make the exponential factor in the output negligible. To do this, t_0/RC should be a large number. The percentage of error in the output of an RC averager, as a function of the ratio of the analysis time t_0 to the time constant RC in the averager, is shown in Fig. 5.22. At small error values (less than 10 percent), the error in an rms value is about 1/2 the error in the mean square value, for a given (T_0/RC) ratio.

It should also be noted that RC averaging must be used if the bandpass filter is to be continuously scanned through the frequency range. When this is the case, a second restriction is placed on the maximum scan rate. (This first is given by Eqs. (5.23) and (5.24).) This restriction is that the scan rate must be slow enough to let the RC averager accurately track changes in the spectrum. The exact value of the scan rate is a function of the error that can be tolerated from this source. In general, the scan rate will be in the following range:

$$\frac{B}{4RC} \leq \lambda \leq \frac{B}{2RC}, \qquad (5.31)$$

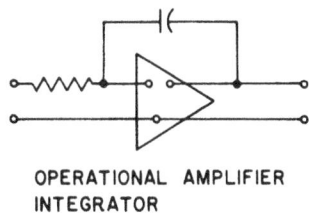

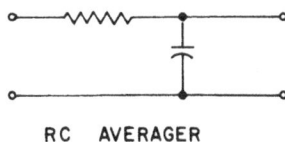

OPERATIONAL AMPLIFIER
INTEGRATOR

RC   AVERAGER

(a)

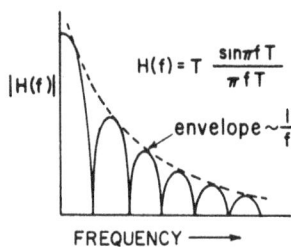

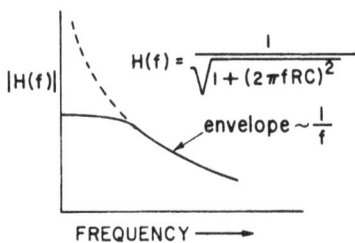

(b)

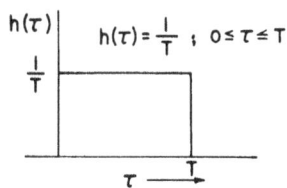

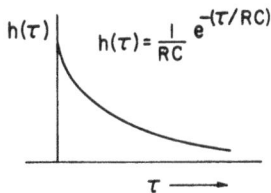

(c)

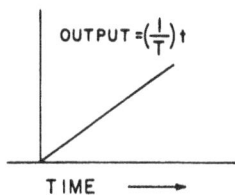

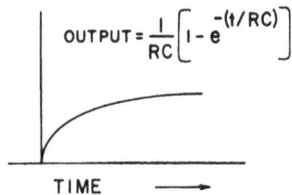

(d)

Fig. 5.21.   Comparison of true integration and RC averaging.

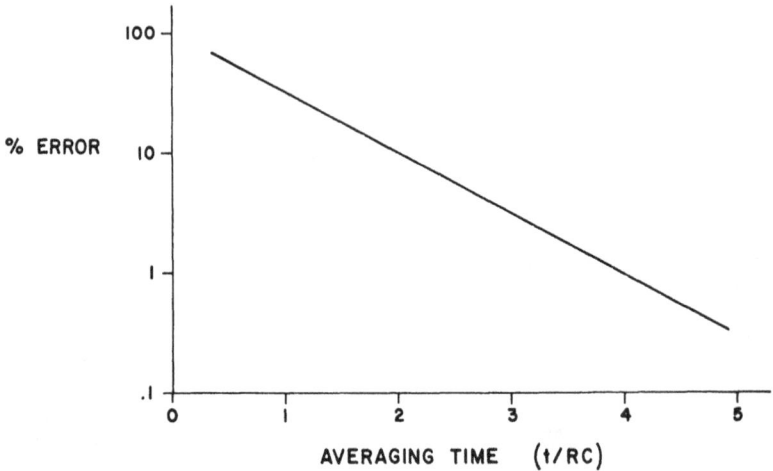

Fig. 5.22.　Error in the mean square value as a function of the averaging time.

where

$$B = \text{the bandwidth of the filter}$$
$$\lambda = \text{the scan rate.}$$

Filter Operation on Transients

Filter-type analog analyzers can be used to compute the Fourier amplitude spectrum without making the transient periodic, if they are of the parallel-filter type. With this type of system, all portions of the frequency range are analyzed simultaneously. When used in this manner, their operation differs somewhat from use on periodic data.

To demonstrate the operation of bandpass filters on a single transient, the time domain response of an idealized, undamped filter will first be calculated. This calculation is performed by means of the convolution integral [26];

$$y(t) = \int_0^t x(\tau)h(t-\tau)d\tau. \tag{5.32}$$

For the undamped filter, the weighting function is

$$h(\tau) = \sin 2\pi f_0 \tau, \tag{5.33}$$

where $f_0 = $ the center frequency of the filter. Then

$$y(t) = \int_0^t x(\tau) \sin (2\pi f_0 t - 2\pi f_0 \tau)d\tau$$
$$= (\sin 2\pi f_0 t) \int_0^t x(\tau) \cos 2\pi f_0 \tau d\tau - (\cos 2\pi f_0 t) \int_0^t x(\tau) \sin 2\pi f_0 \tau d\tau \tag{5.34}$$

$$y(t) = Re\ [X(f_0)] \sin 2\pi f_0 t - Im\ [X(f_0)] \cos 2\pi f_0 t. \qquad (5.35)$$

Thus, the output of the filter contains both the real and imaginary parts of the Fourier transform. To utilize this information, the rms value of the output is computed;

$$\sqrt{\Psi_y^2} = \sqrt{\frac{1}{T}\int_0^T y^2(t)dt}$$

$$= \sqrt{\frac{1}{T}\int_0^T [Re\ [X(f_0)] \sin 2\pi f_0 t - Im\ [X(f_0)] \cos 2\pi f_0 t]^2 dt}.$$

Assume that the transient is terminated, let $T = f_0/4$, and evaluate

$$\sqrt{\Psi_y^2} = \sqrt{Re^2\ [X(f_0)] + Im^2\ [X(f_0)]} = |X(f_0)|. \qquad (5.36)$$

Thus, the rms value of the output is equal to the Fourier amplitude spectrum at f_0. Note that the rms output will be a function of time (running spectrum) until the transient terminates. After this point, the rms output will be constant because of the idealized filter.

The same approach can be used to calculate the response of a practical bandpass filter. The weighting function of a bandpass filter with center frequency f_0 will have the form

$$h(\tau) = g(\tau) \sin 2\pi f_0 \tau, \qquad (5.37)$$

where $g(\tau)$ is some arbitrary function describing the filter characteristics. The output of the filter is calculated as before,

$$y(t) = \int_0^t x(\tau)g(t-\tau) \sin (2\pi f_0 t - 2\pi f_0 \tau)d\tau$$

$$= \sin 2\pi f_0 t \int_0^t x(\tau)g(t-\tau) \cos 2\pi f_0 \tau d\tau - \cos 2\pi f_0 t$$

$$\times \int_0^t x(\tau)g(t-\tau) \sin 2\pi f_0 \tau d\tau. \qquad (5.38)$$

A comparison of Eqs. (5.38) and (5.35) reveals a great deal of similarity, the difference being the factor $g(\tau)$ inside the integrals. If these integrals are designated as

$$Re\ '''[X(f_0)] \text{ and } Im\ '''[X(f_0)], \qquad (5.39)$$

the equation becomes

$$y(t) = Re\ '''[X(f_0)] \sin 2\pi f_0 t - Im\ '''[X(f_0)] \cos 2\pi f_0 t, \qquad (5.40)$$

and the rms value is

$$\sqrt{\overline{\Psi_y^2}} = \left[\frac{1}{T}\int_0^T y^2(t)dt\right]^{1/2} = |\bar{X}(f_0)|^m.$$

(5.41)

This measured spectral value differs from the true value at f_0. The amount by which this measured value deviates from the true value depends on the quantity $g(\tau)$. If $g(\tau)$ does not change significantly over the analysis time interval, the measured value will not differ greatly from the true value. In very general terms, the shorter the duration of the transient and the narrower the bandwidth of the filter, the less the error will be.

5.3   Analog Shock Spectrum Analysis

As with analog Fourier analyzers, analog shock spectrum analyzers can be separated into several classifications. The first grouping is into *parallel* or *scanned* single value categories. With the parallel analyzers, the values of the shock spectrum at all frequencies of interest are obtained simultaneously. With the scanned type of analyzer, the shock spectrum value is obtained at a single frequency, the frequency of the analyzer is stepped, the transient input is repeated, and the value of the shock spectrum is computed at this new frequency. It should be noted that scanning must be performed by stepping rather than by a continuous sweep for this type of analysis.

The other major grouping of shock spectrum analyzers is the separation into either *active* or *passive* analog categories. The active analog shock spectrum computer solves the differential equation

$$\frac{d^2y(t)}{dt} + 2\zeta\omega_n\left[\frac{dy(t)}{dt} - \frac{dx(t)}{dt}\right] + \omega_n^2[y(t) - x(t)] = 0.$$

Active Analog Shock Spectrum Computers

The basic block diagram of an active (direct mathematical model) analog shock spectrum computer was shown in Fig. 5.3. This basic diagram is repeated in Fig. 5.23, with the gains allocated in a better manner and with block diagrams of supplementary functions added.

Prior to the application of the input transient, the outputs of the integrators are clamped to zero. This minimizes the noise in the computing circuitry and is necessary, since zero initial conditions are required in the computations. When the input transient is applied, the timing circuit starts the integrators operating. In addition, this timing circuit governs the time period during which the output of the computing circuitry is connected to the detection circuitry. For measurement of the primary shock spectrum, the output of the computing circuitry is connected to the detectors at or before the start of the input transient.

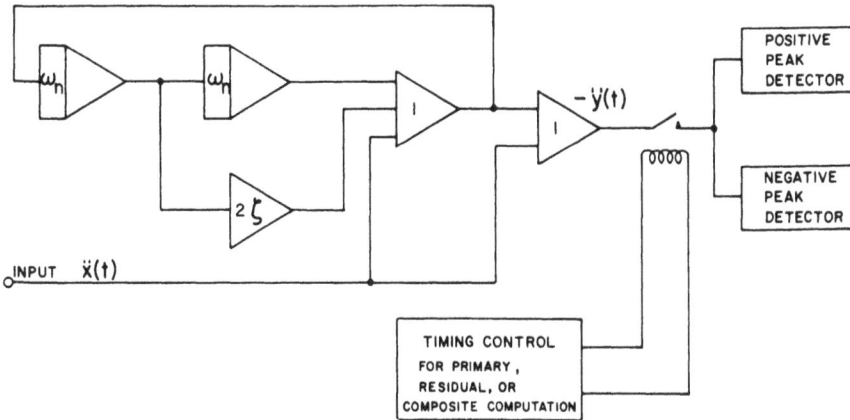

Figure 5.23.   Block diagram of an active analog shock spectrum computer.

It is disconnected in exact coincidence with the end of the input transient. In this way, the measurement is restricted so that the only peaks measured are those occurring during the time interval of the input transient.

For measurement of the residual spectrum, this timing circuit keeps the output from the computing circuit disconnected from the detectors until the *end* of the input transient. Then, in coincidence with the end of the transient, it connects the output of the computing circuit to the detectors. In this way, the measurement time interval is restricted so that the only peaks measured are those occurring after the termination of the input transient.

Since the basic definition of the shock spectrum is the "peak" response of a second order linear system, true peak detectors must be used. Quasi-peak detectors as found in some voltmeters are unsuitable for shock spectrum measurements. A peak, or maximum, is defined as a value of the response whose first time derivative is zero and whose second time derivative is negative. However, peak detectors used in shock spectrum computers do not bother about computing these derivatives, since the exact time of occurrence of the peak is of no concern. (The timing circuits control the data flow and thus handle the primary, residual, or composite spectrum timing.) Peak detectors used in analog shock spectrum analyzers simply measure the highest voltage of a preselected polarity that is applied to their input. The circuit diagram of a simple peak detector is shown in Fig. 5.24.

The charge on the capacitor is bled off by closing the switch. Then the switch is opened for operation. A positive voltage causes the diode to conduct and the charge to be stored on the capacitor. This charge biases the diode so that it will not conduct again until the voltage on the input is greater than the voltage on the capacitor. In this manner, it measures and stores the maximum positive voltage applied to the input.

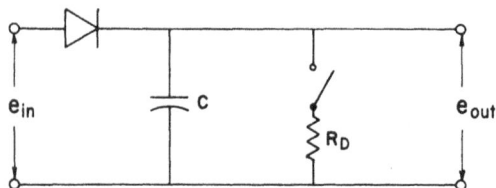

Fig. 5.24. Circuit of a simple peak detector.

This circuit contains two primary sources of error. First, the diode has a finite forward resistance and second, there is a finite shunt resistance across the capacitor. The finite (zero is desired) forward resistance of the diode limits the frequency response of the detector. The frequency response function (assuming conduction of the diode) is approximately that of a simple first order, low-pass filter. The magnitude of the frequency response function can be calculated from

$$|H(f)| = \frac{1}{\sqrt{1 + (2\pi f R_f C)^2}},$$ (5.42)

where R_f = the forward resistance of the diode.

For example, if a 1-percent error can be tolerated at a frequency of 10,000 Hz, then $H(10,000) \geqslant 0.99$. The maximum capacitance value in terms of the forward resistance is found from Eq. (5.42);

$$C \leqslant \frac{2.25}{R_f} \times 10^{-6}.$$

The finite (infinite is desired) shunt resistance across the capacitor causes the charge to bleed off the capacitor. This causes a droop in the voltage stored on the capacitor as shown in Fig. 5.25. The shunt resistance is the parallel combination of the back resistance of the diode, the parasitic shunt resistance of the capacitor, and the input resistance of the device used to read out the stored voltage. The droop can be calculated from

$$e_0(t) = 1 - e^{-t/R_p C},$$ (5.43)

where

$e_0(t)$ = the fractional error from charge being bled off the capacitor

R_p = the shunt resistance across the capacitor (parallel combination of the diode back resistance, capacitor shunt resistance, and following input resistance).

For example, calculate the capacitance required in terms of the shunt resistance across the capacitor to keep the droop to 1-percent over a 1-min time interval;

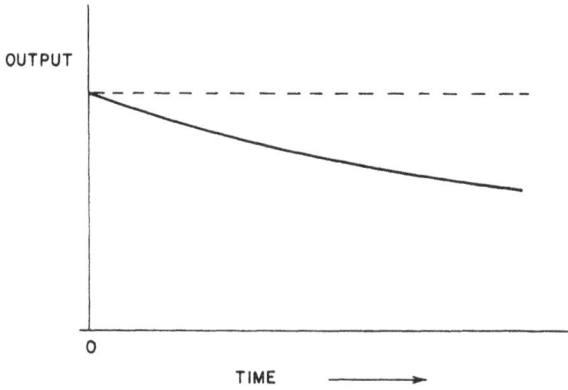

Fig. 5.25. Holding capacitor voltage droop.

$$0.01 = 1 - e^{-t/R_pC}$$

$$t/R_pC = 4.6$$

$$R_pC = \frac{60}{4.6}$$

$$C \geqslant \frac{13}{R_p}.$$

The requirements of Eqs. (5.42) and (5.43) are in direct conflict. The first requirement dictates that the capacitance should be small to minimize the high-frequency rolloff, and the second requirement dictates that the capacitance should be large to prevent the charge from being drained off rapidly. In practice, these requirements limit the range of the capacitor and are reflected as stringent restrictions on the forward resistance of the diode and the total shunt resistance of the circuit across the capacitor.

Generally, this type of shock spectrum computer can be adjusted to have "zero" damping. This adjustment is made by setting the circuit gain so that the response of the computer to a step-function input does not decay or build up over some reasonably large number of cycles of oscillation. Frequently, this adjustment must be performed individually for each frequency point, since minor differences in the gain have a marked effect near the zero damping point.

Passive Analog Shock Spectrum Computers

The second type of shock spectrum computer is based on passive analog computer techniques. The basic block diagram of a passive analog shock spectrum computer was shown in Fig. 5.6. This is the direct electrical analog of the simple base-excited, second order mechanical oscillator. A loop analogy is normally used, since voltage measuring instruments generally disturb the analogy less than current measuring instruments (parasitics from the measuring devices). In Fig. 5.26, the basic block diagram is redrawn and supplementary functions are added. In addition to the basic computing elements, L, R_d, and C, two additional elements are in the computing loop. These are a "negative resistor" and resistor R_1. The latter is simply added to damp out oscillations after the measurement period. The timing control shorts out this resistor when computations start. The negative resistance is included to permit zero damping (or nearly zero damping). Its purpose is to offset the finite resistance in the wiring, the inductor, and the input amplifier stage.

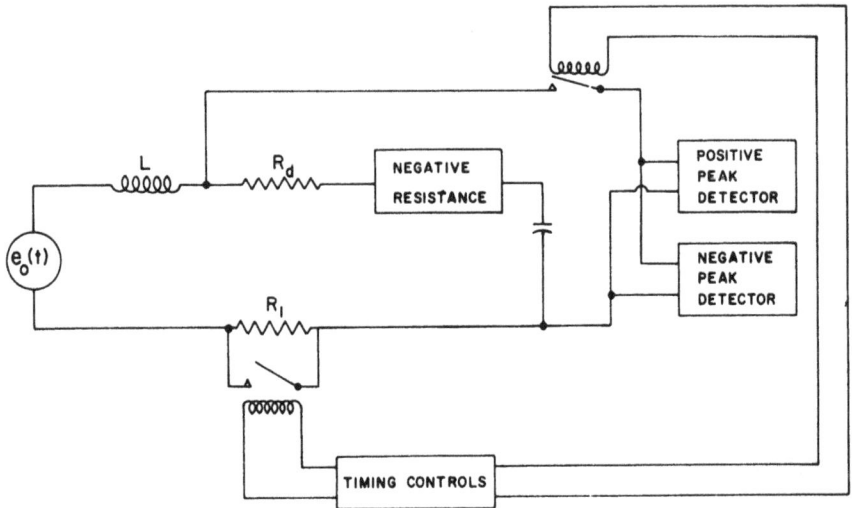

Fig. 5.26.   Block diagram of a passive analog shock spectrum computer.

To illustrate the operation of the negative resistance, the basic computing loop is redrawn in Fig. 5.27. The resistor R_1 is omitted since it plays a supplemental role, and the resistor R_2 has been added. The resistor R_2 represents wiring, inductor, and the input amplifier's output resistance—series parasitic resistance in the loop. The loop equation, in the frequency domain, is

$$E_0(f) = \ddot{Q}(f)L + \dot{Q}(f)R_i - \dot{Q}(f)AR_i + \dot{Q}(f)R_2 + Q(f)/C + \dot{Q}(f)R_d. \qquad (5.44)$$

The current in the loop causes a voltage drop across the input resistor of the amplifier R_i, and this voltage drop is amplified by the amplifier

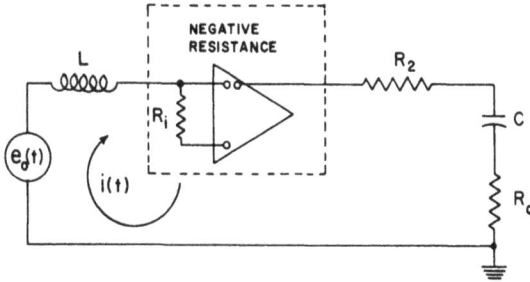

Fig. 5.27.   Loop with negative resistance.

to give a voltage rise. Since the input voltage to the amplifier is QR_1, the output voltage is $A(QR_i)$, where A is the gain of the amplifier. Note that this is a voltage rise rather than a voltage drop. Rewriting Eq. (5.44) gives

$$E_o(f) = \dot{Q}(f)[R_2 - (A-1)R_i] + \ddot{Q}(f)L + \dot{Q}(f)R_d + Q(f)/C. \quad (5.45)$$

By adjusting $(A-1)R_i = R_2$, the parasitic resistances are eliminated and nearly zero damping can be obtained if R_d is set to zero.

It should be noted that the amplifier is floating from ground as shown in Fig. 5.26. This requirement is quite crucial to the operation of this analogy.

Some versions of passive analog shock spectrum computers do not employ negative resistance elements. The primary reason is economic. They, of course, cannot be used to compute undamped shock spectra. Their minimum damping value is limited primarily by the finite resistance of the inductors. Generally, these computers do not contain timing circuits either.

There are two primary sources of errors in passive analog shock spectrum computers. These are (a) parasitics in the inductors and (b) circuit loading by the readout device.

Inductor Parasitic Errors

There are three primary parasitics in inductors. The first is finite dissipation. This is reflected as a resistor in series with the inductor. The dissipation results from three sources:

1. Finite resistance of the wire used in the windings of the inductor,
2. Core losses, and
3. Eddy currents.

This parasitic can be eliminated by the negative resistance approach if the negative resistor is included in the input and inductance leg.

The second parasitic effect is the nonlinearity of the inductance with current. This can be examined by a simplified analysis of the voltage developed across an inductor;

$$e_L(t) = L\frac{di}{dt}$$ (5.46)

or

$$e_L(t) = N\frac{d\Phi}{dt},$$ (5.47)

where

L = inductance
N = number of turns in the inductor
Φ = flux
$\frac{d\Phi}{dt}$ = temporal flux density.

The flux can also be described in terms of its spatial flux density B. In a uniform magnetic field,

$$\Phi = B \cdot A,$$ (5.48)

where A = the area of the field.

Similarly, the spatial flux density can be described in terms of the physical properties of the inductor and the current through the inductor. For a solenoid,

$$B = \mu\,\frac{Ni}{\ell} = \mu H,$$ (5.49)

where

μ = the permeability of the magnetic circuit
ℓ = the length of the solenoid
H = the magnetizing force.

By combining Eqs. (5.47), (5.48), and (5.49), we have

$$e_L(t) = \mu N^2\,\frac{A}{\ell}\frac{di}{dt}.$$ (5.50)

By equating Eqs. (5.46) and (5.50) and solving for the inductance, the following result is obtained:

$$L = \mu\,\frac{N^2 A}{\ell}.$$ (5.51)

Thus, the inductance is a function of the permeability of the magnetic path. The extent of the nonlinearity of the inductor depends on the B vs H curves of the material in the magnetic path. From Eq. (5.49),

$$\mu = \frac{B}{H}.$$

A typical B vs H curve for a ferromagnetic material is shown in Fig. 5.28. Even a cursory examination of this curve points out the great need for care in selecting the core material of the inductors to be used in computing circuits.

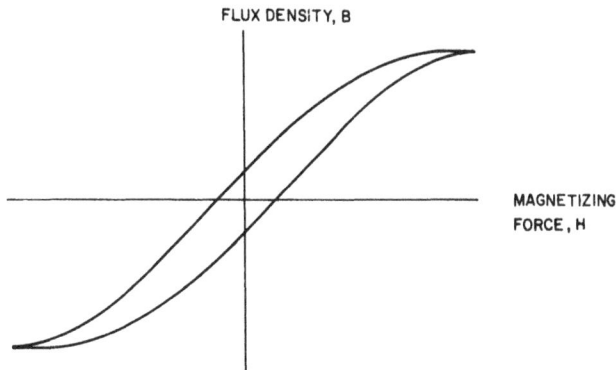

Fig. 5.28.   B vs S curve of a ferromagnetic metal.

The third parasitic effect in inductors is shunt capacitance between the windings. One way of examining the error from this shunt capacitance is to treat the inductance as being frequency-dependent. The inductor, including shunt capacitance C_s, is shown in Fig. 5.29. The total impedance $X_T(f)$ of the inductor is found in the following manner:

$$X_T(f) = \frac{(j2\pi fL)\left(\dfrac{1}{j2\pi fC_s}\right)}{(j2\pi fL)+\left(\dfrac{1}{j2\pi fC_s}\right)}$$

$$= \frac{j2\pi fL}{1-(2\pi f)^2\, LC_s}. \tag{5.52}$$

If it is desired to consider the element as primarily an inductor, corrupted by the interwinding capacitance, the effective inductance L' is

$$L' = \frac{X_T(f)}{j2\pi f} = L\left(\frac{1}{1-(2\pi f)^2\, LC_s}\right) \tag{5.53}$$

or

$$L' = L\left[\frac{1}{1-(f/f_n)^2}\right],$$

where f_n = the frequency at which the inductor resonates solely from the shunt capacitance of the windings.

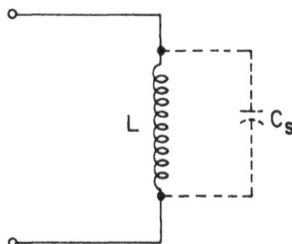

Fig. 5.29.   Inductor showing shunt capacitances.

Ideally, all inductors should be used well below the frequency at which they resonate. While this frequency dependence is totally disastrous in some cases, it is not too bad for shock spectrum computers as long as the resonant frequency of the entire analog circuit is well below the resonant frequency of the inductor alone. This is because the current through the inductor will primarily be at a single frequency. The main problem with the shunt capacitance of the inductor is that it lowers the effective inductance in the inductor.

Loop Loading Errors

The second primary error source in passive analog shock spectrum computers is the loading of the loop circuitry by the readout circuitry. To evaluate the error caused by this loading, the basic loop is redrawn in Fig. 5.30.

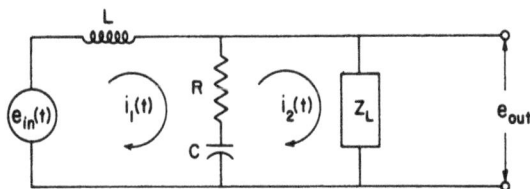

Fig. 5.30. Basic loop with readout loading.

By solving the loop equations in the above figure, the transfer function between the output and input voltages is found to be

$$\left[\frac{E_0(s)}{E_1(s)}\right]_{\text{loaded}} = \frac{R + 1/Cs}{(Ls + R + 1/Cs) + (RLs + L/Cs)/Z_L}. \tag{5.54}$$

The transfer function of the loop without readout loading $(Z_L = \infty)$ is

$$\left[\frac{E_0(s)}{E_1(s)}\right]_{\text{unloaded}} = \frac{R + 1/Cs}{Ls + R + 1/Cs}. \tag{5.55}$$

A comparison of these two equations shows that the result of the loading is to introduce the second term into the denominator. This term must be small in comparison to the first term in the denominator if loading is not to introduce significant error. This leads to the following inequality:

$$\left(\frac{RLs + L/C}{Z_L}\right) \ll Ls + R + 1/(C_s) \tag{5.56}$$

Since this circuit will operate principally at the resonant frequency of the primary loop, Eq. (5.56) can be evaluated at this frequency to determine the primary error. In addition, the magnitudes are the only quantities of interest because of the specific application (a) basically single-frequency operation, and (b) shock spectral analysis does not retain phase information). Under these conditions, Eq. (5.56) evaluated at $f_n = 1/(2\pi\sqrt{LC})$ yields

$$\frac{\sqrt{(L/C)^2 + R^2 L/C}}{|Z_L(f_n)|} \ll R. \tag{5.57}$$

Thus, the magnitude of the load impedance must be quite high at the resonant frequency of the primary loop, as R is always small. If a single readout device is used to read out all frequencies, its impedance must satisfy the requirements of Eq. (5.57) over this entire frequency range.

Other than the error sources just described, there are several other sources of error common to all electronic circuits. The stability of the resonant frequency and damping ratios of the loop depends on the drift characteristics of the elements and the environment in which they are used. Noise is also added to the signal by the electronics. With reasonably careful design, these errors will be minor.

Chapter 6

DIGITAL TECHNIQUES FOR ANALYZING SHOCK DATA

6.1 Digitization of Transient Data

In performing any data analysis digitally, it is first necessary to acquire discrete samples of the data. In most engineering applications, the phenomenon being observed is continuous and the observations are usually recorded in a continuous manner on some medium such as magnetic tape. The continuous data must then be sampled to provide the discrete values required for digital analysis. This sampling process is called digitization.

Digitization consists of two completely independent processes. The first of these is defined as *quantization* and is the procedure whereby one of a discrete set of available numerical values is assigned to the amplitude of the signal being digitized. An example of this is shown in Fig. 6.1. At any point in time, the signal in the example will attain one of the infinitely many values of amplitude possible in the range (0, 4). Since only a discrete subset of amplitude values is available to the quantizing process, say values 0, 1, 2, 3, 4, a decision must be made as to the value assigned. The normal approach is simply to assign the closest level available to the true amplitude. As an example, the true value at time t_1 of the example is approximately 3.8. It would be assigned a value of 4. This selection procedure introduces an error known as *quantizing* error. If the quantizer is working properly, this error will have a zero-mean, uniform probability distribution with a standard deviation of $\sqrt{1/12}\Delta x$, where Δx is the increment between successive quantizing levels.

Since most quantizing devices produce binary outputs so as to be computer-compatible, it is sufficient to define the number of available quantizing levels by specifying the number of binary digits (bits) used. If n is the number of bits available, then there will be 2^n quantizing levels. Similarly,

$$\Delta x = \frac{A}{2^n - 1}, \quad \sigma_\epsilon = \frac{0.29A}{2^n - 1},$$

(6.1)

121

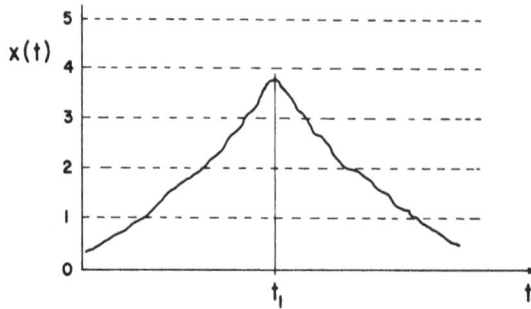

Fig. 6.1   Quantization of a continuous signal.

where A is the amplitude of the signal corresponding to the maximum quantizing level. In most digitizing systems, at least eight and as many as 15 bits are used so that the quantizing error is insignificant. However, care should be taken to utilize as much of the dynamic range of the quantizing system as possible so as to keep the quantizing error at a minimum.

The second process involved in digitization is called *sampling*. Here, one is concerned with the rate at which samples of the data are taken. Digitizing systems will usually supply samples at equally spaced intervals of time. The sampling frequency f_s is then

$$f_s = \frac{1}{\Delta t}, \tag{6.2}$$

where Δt is the time interval. As explained by the sampling theorem [27], only frequency components in the range $(0, f_s/2)$ may be detected in the sampled data. All frequency components greater than $f_s/2$ are folded back or aliased into the acceptable range, as shown in Fig. 6.2:

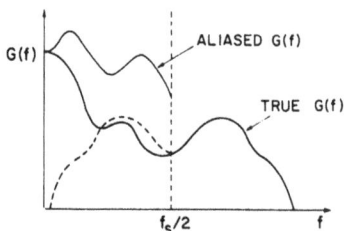

Fig. 6.2.   Aliasing error due to insufficient sampling rate.

Because of the aliasing problem, great care must be taken to see that very little energy exists in the continuous signal beyond the aliasing or Nyquist frequency before digitizing. This is probably the reason for the various rules of thumb specifying the sampling frequency to be used as either five or ten times the highest frequency known to be contained by the data.

The only meaningful way of determining a reasonable sampling frequency is to define the band-limiting properties of the data acquisition system. Since all systems have some upper limits on the frequencies they will pass, it is possible to specify an appropriate sampling frequency at twice this highest frequency. Unfortunately, in many practical applications it is not feasible to sample at the very high rates usually required.

If this is the case, then the data frequency content should be reduced prior to digitizing. The standard method used consists of playing the data through an analog low-pass filter, called an anti-aliasing filter, whose actual passband allows the retention of only the meaningful frequency components while attenuating the high-frequency noise. This procedure may not be effective when dealing with transient data because the filter will produce its own transient response as an output when a transient is supplied as the input. This transient response may have characteristics quite dissimilar from the original data and will tend to ring at the filter's resonant frequency. Also, nonlinear effects can occur. Because of these problems, one must be willing to live with a certain amount of aliasing when performing digital analyses of shock data. The engineer performing this type of analysis should be aware of these potential problem areas and should be prepared to utilize his judgment when confronted with them.

6.2 Classical Digital Fourier Transform Methods

The discrete, finite Fourier transform is defined by

$$X(k\Delta f) = \Delta t \sum_{i=0}^{N-1} x_i e^{-j2\pi k\Delta f i \Delta t}. \tag{6.3}$$

This is simply the discrete analog of Eq. (2.44), where the continuous variables t and f have been replaced by $i\Delta t$ and $k\Delta f$. The natural frequency spacing for the complete Fourier transform, when all possible independent estimates are calculated, is

$$\Delta f = \frac{1}{N\Delta t}, \tag{6.4}$$

so that f_k is defined by

$$f_k = k\Delta f = \frac{k}{N\Delta t}, \qquad k = 0, 1, 2, \ldots, N/2. \tag{6.5}$$

Equation (6.3) may then be rewritten as

$$X_k = \Delta t \sum_{i=0}^{N-1} x_i e^{\frac{-j2\pi i k}{N}}, \qquad k = 0, 1, \ldots, N/2. \tag{6.6}$$

Before the advent of fast Fourier transform (FFT) methods, two approaches were used to evaluate Eq. (6.6).

The first and simplest method used was to evaluate the defining equation. By Euler's formula,

$$e^{\frac{-j2\pi ik}{N}} = \cos\frac{2\pi ik}{N} - j\,\sin\frac{2\pi ik}{N},\tag{6.7}$$

so that

$$X_k = C_k - jQ_k,\tag{6.8}$$

where

$$C_k = \Delta t \sum_{i=0}^{N-1} x_i \cos\frac{2\pi ik}{N}\tag{6.9a}$$

$$Q_k = \Delta t \sum_{i=0}^{N-1} x_i \sin\frac{2\pi ik}{N}.\tag{6.9b}$$

A computer programmer unfamiliar with this type of analysis would proceed to evaluate Eq. (6.9) exactly as written. This would require the calculation of N sines and cosines, the formation of $2N$ products, and the summation of these products for *each* complex value of the Fourier transform. Performed in this manner, the computer time for the complete Fourier transform, ignoring all overhead items such as input-output and initialization, is equal to $(N^2/2)\,(T_{sc} + T_{ma})$, where T_{sc} is the time needed to calculate one value of either the sine or cosine and T_{ma} is the time required to perform one multiplication and one addition. On an IBM 7094 computer, the running time for a reasonable machine language routine would be on the order of $271N^2 \times 10^{-6}$ sec. For reasonable record lengths $(1{,}000 \leq N \leq 10{,}000)$, this procedure requires from 5 min to 7-1/2 hr to evaluate.

It is possible to speed up this procedure by calculating the sines and cosines recursively. This may be done by using the following formulas:

$$\sin(0) = 0$$
$$\sin(\Delta f) = \sin\Delta f \text{ (obtained from sine subroutine)}$$
$$\sin(M+1)\,\Delta f = (2\cos\Delta f)\sin M\Delta f - \sin(M-1)\,\Delta f\tag{6.10a}$$
$$\cos(0) = 1$$
$$\cos(\Delta f) = \cos\Delta f \text{ (obtained from cosine subroutine)}$$
$$\cos(M+1)\,\Delta f = (2\cos\Delta f)\cos M\Delta f - \cos(M-1)\,\Delta f.\tag{6.10b}$$

The values of $\sin\Delta f$ and $\cos\Delta f$ must be calculated by means of an approximation, as is done in most sine–cosine subroutines. However, each additional pair of sines and cosines requires only two multiply–add operations. The running time on an IBM 7094 computer of a machine language routine incorporating the recursion formulas would be on the order of $60 \, N^2 \times 10^{-6}$ sec. For record lengths of $1000 \le N \le 10,000$, the running time is approximately $1 \le T \le 100$ min.

A procedure first derived by Goertzel [28] is reported to be the most efficient of the pre-FFT methods. This method requires the generation of an auxiliary variable U as follows:

$$U_0 = 0$$

$$U_1 = x_{N-1}$$

$$U_i = \left(2 \cos \frac{2\pi k}{N}\right) U_{i-1} - U_{i-2} + x_{N-i}, \quad i = 2, 3, \ldots, N-1. \tag{6.11}$$

Then,

$$C_k = \Delta t \left(\cos \frac{2\pi k}{N}\right) U_{N-1} - U_{N-2} + x_0$$

$$\tag{6.12}$$

$$Q_k = \Delta t \left(\sin \frac{2\pi k}{N}\right) U_{N-1}, \quad k = 0, 1, \ldots, N/2.$$

When this procedure is used and the sines and cosines are generated recursively, a machine language program written for the IBM 7094 will take approximately $20 \, N^2 \times 10^{-6}$ sec of computer time. Again, for reasonable record length ($1000 \le N \le 10,000$) the computer time will be in the neighborhood of 20 sec to 33 min. While this is a saving of a factor of three in computer time over the previous method, note that the running time is still of the order N^2. For large volumes of data, computer times of this order are generally unacceptable. The end result has been that very little transient data have been analyzed in this manner. Instead, transients have usually been analyzed either by inspection of the excitation itself or by shock spectrum analysis.

6.3 Fast Fourier Transform Methods

In 1965, a paper by Cooley and Tukey [29] was published describing a "new" method for calculating Fourier series or Fourier transforms. The basis for this method has since been traced back to at least 1928. Recently, considerable work on this subject has appeared in print [30–33], and two major versions of the algorithm have been defined. These are presently termed the Cooley-Tukey (C-T) algorithm and the Sande-Tukey (S-T) algorithm.

The advantages of the FFT are twofold. The number of actual arithmetic operations is reduced drastically, causing increases in speed of several orders of magnitude for reasonable record lengths. Also, because of the fewer operations performed, truncation and roundoff errors are reduced, producing a more accurate result.

The major restriction on the use of the FFT method is that the number of samples N be a highly composite number. That is

$$N = r_1 \cdot r_2 \cdot r_3 \ldots \cdot r_m,$$

where r_i is a non-unity factor of N. The number of complex multiply–add operations required by the FFT method is proportional to $N \Sigma r_i$ instead of N^2. In particular, for $N = 2^p$, the number of operations is approximately $2Np$.

In essence, the FFT algorithms are methods for factoring the Fourier transform of order N into a series of transforms, each of which is of order r_i. This can be seen more easily from the following example:

The Fourier transform of $x(t)$ is defined by

$$X(k) = \Delta t \sum_{i=0}^{N-1} x_i e^{\frac{-j2\pi ik}{N}}, \qquad k = 0, 1, \ldots, N/2. \tag{6.13}$$

A change in notation is appropriate here. Let

$$W = e^{\frac{-j2\pi}{N}}. \tag{6.14}$$

Then

$$X(k) = \Delta t \sum_{i=0}^{N-1} x_i W^{ik}. \tag{6.15}$$

If $N = A \cdot B$, then the two indices i and k may be rewritten

$$i = (b + aB), \text{ time index}$$
$$k = (c + dA), \text{ frequency index}, \tag{6.16}$$

where

$$a, c = 0, 1, \ldots, A - 1$$
$$b, d = 0, 1 \ldots, B - 1.$$

Substituting these indices into Eq. (6.15) produces

$$X(c + dA) = \sum_{b=0}^{B-1} \sum_{a=0}^{A-1} x(b + aB) W^{(b+aB)(c+dA)}, \tag{6.17}$$

where the Δt scale factor has been omitted to simplify the equation.

Expanding the exponents of W, the following is obtained:

$$W^{(b+aB)(c+dA)} = W^{bc} W^{bdA} W^{acB} W^{adAB}. \tag{6.18}$$

Also,

$$W^{adAB} = e^{-\frac{j2\pi adAB}{AB}} = [e^{-j2\pi}] \, ad = 1, \tag{6.19}$$

because a and d are both integers and $e^{-j2\pi}$ raised to any integer power is unity. Eq. (6.17) can now be rewritten as

$$X(c+dA) = \sum_{b=0}^{B-1} \left\{ W^{bdA} \left[W^{bc} \sum_{a=0}^{A-1} x(b+aB) \, W^{acB} \right] \right\}. \qquad (6.20)$$

The inner summation is recognizable as a discrete Fourier transform of the decimated data sequence $x(b+aB)$ whose length is $N/B=A$. Simplify the notation and let

$$u_b(a) = x(b+aB), \qquad a=0, 1, \ldots, A-1; \, b=0, 1, \ldots, B-1. \quad (6.21)$$

Define

$$U_b(c) = \sum_{a=0}^{A-1} u_b(a) \, W^{acB}, \qquad b=0, 1, \ldots, B-1. \qquad (6.22)$$

B of these A-length transforms must be calculated. Define

$$v_c(b) = U_b(c) \, W^{bc}. \qquad (6.23)$$

Then Eq. (6.20) may be rewritten as

$$X(c+dA) = \sum_{b=0}^{B-1} v_c(b) \, W^{bdA}, \qquad d=0, 1, \ldots, B-1; \, c=0, 1, \ldots, A-1,$$

$$(6.24)$$

which is nothing more than A Fourier transforms, each of length B. Therefore, the original Fourier transform has been broken down into two stages:
- Computation of B Fourier transforms of length A
- Computation of A Fourier transforms of length B.

The number of complex multiply–add operations required is as follows:

$$A \cdot B^2 + B \cdot A^2 = AB \, (A+B). \qquad (6.25)$$

While N may be composed of any factors, either prime or nonprime, the most efficient factors are either 2 or 4. This is because the intermediate transforms will be of length 2 or 4 and the exponentials required for such transforms will have values of $+1$ and -1 or $+1$, $+j$, -1, $-j$. This allows the replacement of complex multiplications by complex additions and subtractions, which makes for additional time savings. Also, the programming becomes somewhat simpler when all the factors of N are identical.

The algorithms as described in the literature define a recursive approach in the calculation of the intermediate transform values. In effect, the recursion process is a simple method for calculating the $(i+1)$st intermediate transform from the ith.

The description of the recursion equations to follow will be restricted to the power-of-two versions only. First, the time and frequency indices i and k must be written in binary notation, as

$$i = i_{p-1}2^{p-1} + i_{p-2}2^{p-2} + i_{p-3}2^{p-3} \ldots + i_0 = (i_{p-1}, \ldots, i_0) \qquad (6.26)$$

and

$$k = k_{p-1}2^{p-1} + k_{p-2}2^{p-2} + k_{p-3}2^{p-3} \ldots + k_0 = (k_{p-1}, \ldots, k_0),$$

where each component of i and k can take on only the values 0 or 1. These indices are usually written in a positional form with the powers of two omitted as shown by the extreme right-hand sides of the definitions. As will be seen, the binary indices correspond to the actual computer memory addresses of the data samples being processed.

The describing equation for the Fourier transform may now be written in a manner analogous to Eq. (6.17) as

$$X(k_{p-1}, k_{p-2}, \ldots, k_0) = \sum_{i_0=1}^{1} \sum_{i_1=0}^{1} \sum_{i_2=0}^{1} \cdots \sum_{i_{p-1}=0}^{1} x(i_{p-1}, i_{p-2}, \ldots, i_0) W^{ik}. \qquad (6.27)$$

Expanding the complex exponential W^{ik} as before allows the factoring of integral powers of $e^{-j2\pi}$. This can be seen most clearly by considering the product of only the $k_{\ell-1}$ index with the complete i index:

$$W^{k_{\ell-1}(i_{p-1}, i_{p-2}, \ldots, i_0)} = W^{k_{\ell-1}(i_{p-1}, i_{p-2}, \ldots, i_{p-\ell+1})}$$

$$W^{k_{\ell-1}(i_{p-\ell}, i_{p-\ell-1}, \ldots, i_0)}. \qquad (6.28)$$

The first term on the right-hand side of Eq. (6.28) now contains only integral powers of the complex exponential, since the product of $k_{\ell-1}$ and each of the i indices is of the form 2^{p+M}:

$$W^{ik} = e^{-\frac{j2\pi ik}{N}} = e^{-\frac{j2\pi ik}{2^p}} = e^{-\frac{j2\pi 2^{p+M}}{2^p}} = e^{-j2\pi 2^M}. \qquad (6.29)$$

Equation (6.27) may now be rewritten in recursive form. Consider the innermost sum performed over $i_{p-1} = 0, 1$. Using the notation A_ℓ to denote the ℓth intermediate transform, the innermost sum may be written as

$$A_1(k_0, i_{p-2}, i_{p-3}, i_{p-4}, \ldots, i_0) = \sum_{i_{p-1}=0}^{1} x(i_{p-1}, \ldots, i_0) W^{k_0 i_{p-1}} \qquad (6.30)$$

The next stage in the recursion is obtained from the first by

$$A_2(k_0, k_1, i_{p-3}, i_{p-4}, \ldots, i_0) = \sum_{i_{p-2}=0}^{1} A_1(k_0, i_{p-2}, i_{p-3}, \ldots, i_0) W^{(k_1+k_0)i_{p-2}} \qquad (6.31)$$

and in general,

$$A_\ell(k_0, k_1, \ldots, k_{\ell-1}, i_{p-\ell-1}, \ldots, i_0) = \sum_{i_{p-\ell}=0}^{1} A_{\ell-1}(k_0, k_1, \ldots,$$

$$k_{\ell-2}, i_{p-\ell}, \ldots, i_0) W^{(k_{\ell-1}+k_{\ell-2}+\cdots k_0)i_{p-\ell}}. \qquad (6.32)$$

The pth application of this recursion produces the required Fourier transform. However, the index of the transform is in reverse order, going from the least significant to the most significant digit instead of vice-versa. Thus, the Fourier transform values must be rearranged before the process is completed. This rearrangement is termed a bit reversal.

This version of the FFT is the C–T algorithm. The S–T algorithm is similar, but the roles of the two indices in the exponents are interchanged. The relation for the S–T algorithm is

$$A_\ell(k_0, k_1, \ldots, k_{\ell-1}, i_{p-\ell-1}, \ldots, i_0) = \sum_{i_{p-\ell}=0}^{1} A_{\ell-1}(k_0, k_1, \ldots, k_{\ell-2},$$

$$i_{p-\ell}, \ldots, i_0) W^{k_{\ell-1}(i_{p-\ell}+i_{p-\ell-1}\cdots, i_0)}. \qquad (6.33)$$

The forms of these algorithms which are actually used for programming are given below.

a. The C–T algorithm

$$A_\ell(k_0, k_1, \ldots, k_{\ell-2}, 0, i_{p-\ell-1}, \ldots, i_0) = [A_{\ell-1}(k_0, \ldots, k_{\ell-2}, 0, i_{p-\ell-1}, \ldots, i_0)$$

$$+ A_{\ell-1}(k_0, k_1, \ldots, k_{\ell-2}, 1, i_{p-\ell-1}, \ldots, i_0)] W^{(k_{\ell-2}+\cdots+k_0)i_{p-\ell-1}}. \qquad (6.34)$$

$$A_\ell(k_0, k_1, \ldots, k_{\ell-2}, 1, i_{p-\ell-1}, \ldots, i_0)$$
$$= [A_{\ell-1}(k_0, k_1, \ldots, k_{\ell-2}, 0, i_{p-\ell-1}, \ldots, i_0)$$
$$- A_{\ell-1}(k_0, k_1, \ldots, k_{\ell-2}, 1, i_{p-\ell-1}, \ldots, 1_0)] W^{(k_{\ell-2}+\cdots+k_0)i_{p-\ell-1}}.$$

b. The S–T algorithm:

$$A_\ell(k_0, k_1, \ldots, k_{\ell-2}, 0, i_{p-\ell-1}, \ldots, i_0)$$
$$= [A_{\ell-1}(k_0, k_1, \ldots, k_{\ell-2}\, 0, i_{p-\ell-1}, \ldots, i_0)$$
$$+ A_{\ell-1}(k_0, k_1, \ldots, k_{\ell-2}, 1, i_{p-\ell-1}, \ldots, i_0)] W^{(i_{p-\ell-1}\cdot i_{p-\ell-2}\cdots i_0)k_{\ell-1}}$$
$$\qquad (6.35)$$

$$A_\ell(k_0, k_1, \ldots, k_{\ell-2}, 1, i_{p-\ell-1}, \ldots, i_0)$$
$$= [A_{\ell-1}(k_0, k_1, \ldots, k_{\ell-2}, 0, i_{p-\ell-1}, \ldots, i_0)$$
$$- A_{\ell-1}(k_0, k_1, \ldots, k_{\ell-2}, 1, i_{p-\ell-1}, \ldots, i_0)] W^{(i_{p-\ell-1}\cdot i_{p-\ell-2}\cdots i_0)k_{\ell-1}}.$$

The formulas given above show the simple patterns which may be implemented in the programming. Two values of the intermediate transform are calculated at one time. These two values differ in index by a factor of two in the $(p - \ell)$ th position. If the original data were stored in consecutive computer memory locations, then the pair of intermediate transforms will appear in locations $2^{p-\ell}$ apart. Also, the two values are almost identical; one is a function of the sum of two previous values, while the other is the difference of these values. Because of these patterns, the implementation of an FTT algorithm is almost trivial.

Another item of interest is the fact that only $2N$ sines and cosines are required rather than the N^2 values required by earlier methods. Because of this reduced number, it is practical to calculate each sine and cosine independently rather than recursively. This reduces the computational error, since roundoff and truncation errors usually begin to show up in the later stages of recursive sine–cosine generation techniques.

The final procedure required in most versions of the FFT is the bit reversal to unscramble the final transforms. The reversed index may be computed in one of two ways:

1. By maintaining a counter in which the increment is added to the most significant digit rather than the least, as is usual. This presents a few difficulties in the handling of carries, since the carry must be to the right rather than to the left.

2. In a recursive manner, from the preceding value of the reversed-bit index. To accomplish this,

(a) Search the leading digits of the preceding index for a zero. Set all leading nonzero bits to zero.

(b) Set the first zero bit to a one. All other less significant bits maintain their previous values.

The bit reversal can just as easily be implemented prior to transformation. The identical process may be used to scramble the time series before the FFT algorithm is applied. The results of the transformation will then occur in the correct order.

Several other facets of the FFT are of major interest. Since the procedure transforms a complex sequence of data, it is necessary to fill in the imaginary part of a time history with zeros. However, if transforms of several data sequences are required, pairs of transforms may be obtained simultaneously by forming a complex time series composed of two real time series as follows:

Given

$$x_i = x(i\Delta t), \qquad y_i = y(i\Delta t),$$

then form

$$z_i = x_i + j y_i, \qquad i = 0, 1, \ldots, N-1. \tag{6.36}$$

The Fourier transform of z is obtained in the usual manner and then unscrambled to produce the two transforms

and

$$X_k = \frac{Z_k + Z^*_{N-k}}{2}$$

$$Y_k = \frac{Z_k - Z^*_{N-k}}{2}, \quad k = 0, 1, \ldots, N/2.$$

(6.37)

Limitations in FFT Methods

The basic limitations of the FFT method are twofold. First, the number of samples must be a highly composite number and, for most presently programmed versions of the FFT, a power of two. This limitation can be overcome simply by either truncating the time history at some convenient point or by adding zero values until the required record length is obtained. In either case, no appreciable effects on the resultant transform will be apparent. The padding with zeros will simply introduce some additional values of the transform. These will not be statistically independent, as are those values calculated at the natural frequencies. In fact, adding zeros to the time history is a good interpolation procedure, since it corresponds to the error-free $(\sin x)/x$ interpolator defined by Shannon's sampling theorem [27].

The second limitation of the method is that the entire data sequence must normally be available before the transformation can be performed. Since the transform is performed in place with the intermediate values being placed on top of preceding values, the number of samples which may be processed is limited by the size of the computer memory. With binary computers, sample sizes of approximately 1/4 the computer memory capacity may be processed. This is because binary computers have memory sizes equal to some power of two. Since the FFT is a complex transformation, two memory locations are required for each sample of the time series, so that for $N = 2^\nu$, $2^{\nu+1}$ actual locations are needed. If 2^ν is greater than 1/4 of the total computer memory, then it would have to be equal to at least 1/2 of the total memory, so that $2^{\nu+1}$ (the number of memory locations required) would be equal to or greater than the total memory.

However, by a very simple procedure it is possible to compute transforms of record lengths twice the size of the normal maximum. The technique is as follows:

1. Sort the data samples into two subsets, one containing the even-indexed points and the other the odd-indexed points.

2. Perform the FFT on each subset.

3. Combine the resultant transforms to produce the double-length transforms.

In equation form, this process is the following:

$$a_k = x_{2i}$$

$$b_k = x_{2i+1}, \qquad i = 0, 1, \ldots, N-1; \ k = 0, 1, \ldots, N/2-1 \tag{6.38}$$

$$A_k = \sum_{i=0}^{N/2-1} a_k (W^2)^{ik}$$

$$B_k = \sum_{i=0}^{N/2-1} b_k (W^2)^{ik}, \qquad k = 0, 1, \ldots, N/2-1 \tag{6.39}$$

$$X_k = A_k + B_k W^k$$

$$X_{k+N/2} = A_k - B_k W^k, \qquad k = 0, 1, \ldots, N/2-1. \tag{6.40}$$

This procedure may be repeated to allow successive doubling of the number of data points indefinitely. To implement this technique efficiently, the computer used should have some random-access auxiliary memory such as a disk or drum for intermediate storage.

6.4 Shock Spectrum Analysis Methods

Shock spectra are calculated digitally in a manner similar to the analog procedures discussed earlier. The response histories of a series of single degree-of-freedom systems to the given excitation are calculated at specified natural frequencies and damping ratios. The peaks of these response histories are detected and recorded as a function of their natural frequencies and damping ratios to produce the shock spectrum.

However, the various procedures used in calculating the response histories and detecting response peaks are quite different from the analog techniques. The response history computations in present use consist of

1. Direct numerical integration of the Duhamel integral
2. Recursive integration of the Duhamel integral
3. Convolutional filtering by means of the single degree-of-freedom system unit impulse response
4. Recursive filtering procedures.

All of these techniques are discussed in detail in following sections.

Digital peak detection is an area of some interest and is a major source of error. Since the response history is sampled rather than continuous, the probability of observing the actual maximum is remote. Interpolation schemes of various types are in use to perform the peak detection and evaluation. A more complete discussion of this problem appears later in this chapter.

6.5 Response History Computation via Integration

The initial procedure implemented digitally for computing the response history of a single degree-of-freedom system was the numerical integration of the Duhamel or superposition integral. As described in Chapter 4, this procedure involves the convolution of the exciting function with the unit impulse response of a single degree-of-freedom system. In describing this procedure, the foundation-excited form of the system will be used. The differential equation of the motion of the mass as derived in Chapter 4, Eq. (4.24), is then just

$$\ddot{y}(t) + 2\zeta\omega_n[\dot{y}(t) - \dot{x}(t)] + \omega_n^2[y(t) - x(t)] = 0, \tag{6.41}$$

with a general solution of

$$\xi(t) = \xi_0 e^{-\zeta\omega_n t}\left(\cos \omega_d t + \frac{\zeta}{\sqrt{1-\zeta^2}} \sin \omega t_d\right) + \frac{\dot{\xi}_0}{\omega_d} e^{-\zeta\omega_n t} \sin \omega_d t$$

$$- \frac{1}{\omega_d}\int_0^t \ddot{x}(\tau)e^{-\zeta\omega_n(t-\tau)} \sin \omega_d(t-\tau)d\tau. \tag{6.42}$$

where ξ_0 and $\dot{\xi}_0$ are the initial relative displacement and velocity, respectively. If the initial conditions are zero, as they usually are in practice, then the equation to be solved is simply

$$\xi(t) = -\frac{1}{\omega_d}\int_0^t \ddot{x}(\tau)e^{-\zeta\omega_n(t-\tau)} \sin \omega_d(t-\tau)d\tau. \tag{6.43}$$

A similar equation may be derived for the relative velocity $\dot{\xi}(t)$;

$$\dot{\xi}(t) = -\int_0^t \ddot{x}(\tau)e^{-\zeta\omega_n(t-\tau)}\left[\cos \omega_d(t-\tau) - \frac{\zeta}{\sqrt{1-\zeta^2}} \sin \omega_d(t-\tau)\right]d\tau. \tag{6.44}$$

which may be rewritten

$$\dot{\xi}(t) = -\int_0^t \ddot{x}(\tau)e^{-\zeta\omega_n(t-\tau)} \cos \omega_d(t-\tau)d\tau - \zeta\omega_n\xi(t). \tag{6.45}$$

Also,

$$\ddot{\xi}(t) = -\int_0^t \ddot{x}(\tau)\omega_n e^{-\zeta\omega_n(t-\tau)}\left[\frac{2\zeta^2-1}{\sqrt{1-\zeta^2}} \sin \omega_d(t-\tau) - 2\zeta \cos \omega_d(t-\tau)\right]d\tau. \tag{6.46}$$

Similarly, this equation may be rewritten in terms of $\xi(t)$,

$$\ddot{\xi}(t) = 2\zeta\omega_n\int_0^t \ddot{x}(\tau)e^{-\zeta\omega_n(t-\tau)} \cos \omega_d(t-\tau)d\tau + \omega_n(2\zeta^2-1)\xi(t). \tag{6.47}$$

In any case, it may be seen that the solution of these equations requires the integration of a damped sinusoid or cosinusoid multiplied by the time history of the excitation. The direct translation of these equations into the digital domain requires only the replacement of the continuous variable $x(t)$ by the discrete $x_i = x(i\Delta t)$ and of the integral by a summation. The discrete form of Eq. (6.43) may now be written as

$$\xi_i = -\frac{\Delta t}{\omega_d} \sum_{k=0}^{i} \ddot{x}_k e^{-\zeta\omega_n \Delta t(i-k)} \sin\left[\omega_d \Delta t(i-k)\right]. \tag{6.48}$$

The accuracy of the numerical integration performed in this manner may leave much to be desired. In essence, the continuous excitation $\ddot{x}(t)$ is replaced by a series of rectangles of width Δt as shown in Fig. 6.3. If the ratio of the sampling frequency f_s to the highest frequency component contained in the excitation f_H is sufficiently large, say $\geqslant 20$, then this integration procedure will produce negligible error in the response history. However, such sampling frequencies increase the computer time drastically.

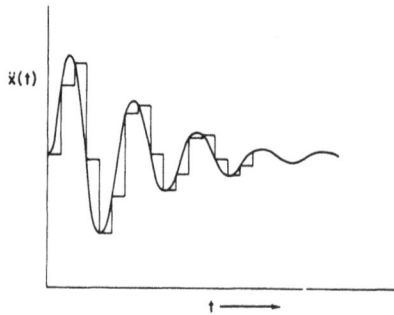

Fig. 6.3. Numerical integration of an excitation time history.

As an example, consider a complex shock pulse containing frequency components of interest in the range 0 to 10 kHz. If this pulse is sampled at 20 times the highest frequency, the sampling frequency will be 200,000 samples/sec. Typical shock pulses will usually take 10 to 100 msec (0.01 to 0.1 sec) to decay. This means that as many as 20,000 data samples might be required to define the excitation. A reasonable estimate for the time required to calculate one integration step on the IBM 7094 is 540 μsec. With an average of $20,000/2 = 10,000$ integration steps required for each value of the response history, this means that 5.4 sec of computer time will be expended. Considering the number of values of the response history to be calculated (typically 5 to 10 times per cycle of the resonant frequency) and the number of natural frequencies usually required to give reasonable resolution to the shock spectrum, it is obvious that the computer time required is prohibitive.

Techniques for speeding this computation have been implemented in past computer programs. These include the following:

1. Recursive generation of the exponential sines and cosines. Since the values of the excitation $x(t)$ are normally available at equal increments of time Δt it is necessary to compute only initial values of the exponential sines and cosines $e^{-\zeta \omega_n t} \sin \omega_d t$ and $e^{-\zeta \omega_n t} \cos \omega_d t$ and similar values for the increment Δt; $e^{-\zeta \omega_n \Delta t} \sin \omega_d \Delta t$ and $e^{-\zeta \omega_n \Delta t} \cos \omega_d \Delta t$. Recursive relations based on the formulas for the sines and cosines of the difference of two angles and for the difference of two exponents can be used;

$$\sin\left[\omega_d(t_i - \Delta t)\right] = \sin \omega_d t_{i-1} = \sin \omega_d t_i \cos \omega_d \Delta t - \cos \omega_d t_i \sin \omega_d \Delta t \quad (6.49)$$

$$\cos \omega_d t_{i-1} = \cos \omega_d t_i \cos \omega_d \Delta t + \sin \omega_d t_i \sin \omega_d \Delta t \quad (6.50)$$

$$e^{-\zeta \omega_n t_{i-1}} = e^{-\zeta \omega_n t_i} e^{\zeta \omega_n \Delta t} \quad (6.51)$$

Combining these formulas yields

$$e^{-\zeta \omega_n t_{i-1}} \sin \omega_d t_{i-1} = (e^{-\zeta \omega_n t_i} e^{\zeta \omega_n \Delta t})(\sin \omega_d t_i \cos \omega_d \Delta t - \cos \omega_d t_i \sin \omega_d \Delta t)$$

$$= (e^{\zeta \omega_n \Delta t} \cos \omega_d \Delta t)(e^{-\zeta \omega_n t_i} \sin \omega_d t_i)$$

$$- (e^{\zeta \omega_n \Delta t} \sin \omega_d \Delta t)(e^{-\zeta \omega_n t_i} \cos \omega_d t_i). \quad (6.52)$$

The first of the two factors in each term is simply the incremental value of the exponential sine or cosine, while the second factor is the preceding value of the exponential sine or cosine. Done in this manner, only four multiplies and two adds are required to generate each term of the single degree-of-freedom, unit-impulse response function.

2. Limiting the number of response points calculated. If only the maximax or maximum positive and maximum negative shock spectra are to be calculated, it is sufficient to compute the response histories over only a few cycles of the frequencies of interest. This is because the peak response usually occurs within 2 or 3 response cycles of the peak excitation. To implement this limitation procedure, it is only necessary to specify the record length to be used in the computations as some number of periods of the natural frequencies required.

3. Limiting the number of values per response period. Another limitation which can be made is to specify the maximum number of response values required per cycle of the resonant frequency. It is not usually necessary to maintain more than 8 to 10 response samples per cycle to obtain accurate estimates of the response peaks. This limitation is especially effective in reducing computer time in a broadband analysis. Because of the high-frequency data content, high sampling rates are required. If a response were to be calculated for each sample of the excitation, many more values of the response than necessary would be computed for the lower and middle frequency range.

6.6   Response History Computation via Recursion Formulas

Another procedure similar to the direct integration of the Duhamel integral, as described in the preceding section, but much more efficient is a recursion due to O'Hara [34]. A variant of this method is also reported by Gertel [19].

This technique utilizes Eqs. (6.43) and (6.45) which define the relative deflection and relative velocity of a single degree-of-freedom system as a function of the initial relative displacement and velocity and also of the convolution of the excitation with the system response function. It is easily seen that if the relative displacement and velocity are known at some time t_1, where t_1 is later than the time of initiation of the shock, then these values could be used as new initial values in determining the response time history. Therefore, if the Duhamel integral can be integrated properly over some time interval, then the response values $\xi(t_1)$ and $\dot{\xi}(t_1)$ computed in this way can be used as new starting values to continue the solution process.

In general, the exciting function is available at equally spaced intervals of time. While the equal spacing is not a requirement in the recursion, it does simplify both the notation and the programming of the method.

Consider an acceleration record as shown in Fig. 6.4 in which discrete samples are taken at the equally spaced points t_i. The determination of the response values ξ_i and $\dot{\xi}_i$ resolves itself into one of numerically integrating the Duhamel integral over the intervals

$$\langle t_0, t_1 \rangle, \langle t_1, t_2 \rangle, \ldots \langle t_k, t_{k+1} \rangle, \ldots \langle t_{n-1}, t_n \rangle.$$

To perform this numerical integration, it is necessary to determine an analytic expression which approximates the quantity to be integrated. This boils down to approximating the exciting function by some simple form, since the remainder of the integrand is simply a sine or cosine.

A good approximation in most cases is that of a parabola such that

$$\hat{\ddot{x}} = a_0 + a_1 t_i + a_2 t_i^2, \tag{6.53}$$

where the coefficients a_0, a_1, and a_2 may be determined from the samples of the excitation in the vicinity of t_i. In particular, by utilizing finite difference methods, the parabolic approximation may be written as

$$\hat{\ddot{x}}_i = \ddot{x}_i + \frac{S_i t_i}{\Delta t} + \frac{S_{i-1}^2}{2} \left(\frac{t_i^2}{\Delta t^2} - \frac{t_i}{\Delta t} \right), \tag{6.54}$$

where

$$S_i = \ddot{x}_{i+1} - \ddot{x}_i \quad \text{(first forward difference)}$$

$$S_{i-1}^2 = S_i - S_{i-1} = \ddot{x}_{i+1} - 2\ddot{x}_i + \ddot{x}_{i-1} \quad \text{(second forward difference)}.$$

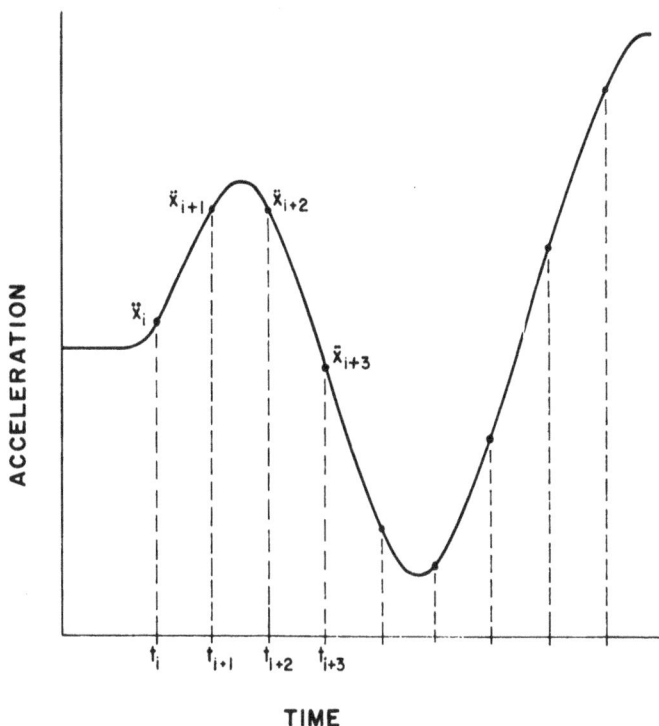

Fig. 6.4.   Digitized acceleration record.

Utilizing the parabolic approximation of Eq. (6.54) for the exciting function, an explicit solution of the Duhamel integral is obtained:

$$\int_{t_i}^{t_{i+1}} x(\tau)e^{-\zeta\omega_n(t-\tau)}\sin \omega_d(t-\tau)d\tau$$

$$= \frac{-\ddot{x}_i}{\omega_n^2}\left[1 - e^{-\zeta\omega_n\Delta t}\left(\cos \omega_d\Delta t + \frac{\zeta}{\sqrt{1-\zeta^2}}\sin \omega_d\Delta t\right)\right]$$

$$\frac{-S_i}{\omega_n^2}\left[1 - \frac{2\zeta}{\omega_n\Delta t}(1 - e^{-\zeta\omega_n\Delta t}\cos \omega_d\Delta t) - \frac{(1-2\zeta^2)e^{-\zeta\omega_n\Delta t}\sin \omega_d\Delta t}{\omega_d\Delta t}\right]$$

$$-\frac{S_{i-1}^2}{2\omega_n^2}\left[-\frac{4\zeta}{\omega_n\Delta t} - \left(\frac{2(1-4\zeta^2)}{\omega_n^2\Delta t^2} - \frac{2\zeta}{\omega\Delta t}\right)(1 - e^{-\zeta\omega_n\Delta t}\cos \omega_d\Delta t)\right.$$

$$\left. + \left(\frac{1-2\zeta^2}{\omega_n\Delta t} + \frac{2\zeta(3-4\zeta^2)}{\omega_n^2\Delta t^2}\right)\frac{e^{-\zeta\omega_n\Delta t}\sin \omega_d\Delta t}{\sqrt{1-\zeta^2}}\right]. \qquad (6.55)$$

In a similar manner, an explicit solution for the cosine integral may be obtained. If these substitutions are made in the discrete forms of Eqs. (4.21) and (4.27), the resultant equations may be used recursively to obtain the $(k+1)$st values of the relative responses from the kth values;

$$\xi_{k+1} = B_1\xi_k + B_2\dot{\xi}_k + B_3\ddot{x}_k + B_4S_k + B_5S_{k-1}^2 \tag{6.56}$$

$$\frac{\dot{\xi}_{k+1}}{\omega_n} = B_6\xi_k + B_7\dot{\xi}_k + B_8\ddot{x}_k + B_9S_k + B_{10}S_{k-1}^2, \tag{6.57}$$

where

$$B_1 = e^{-\zeta\omega_n\Delta t}\left(\cos \omega_d\Delta t + \frac{\zeta}{\sqrt{1-\zeta^2}}\sin \omega_d\Delta t\right) \tag{6.58}$$

$$B_2 = \frac{e^{-\zeta\omega_n\Delta t}}{\omega_d}\sin \omega_d\Delta t \tag{6.59}$$

$$B_3 = \frac{1}{\omega_n^2}(1-B_1) \tag{6.60}$$

$$B_4 = \frac{1}{\omega_n^2}\left[1 - \frac{2\zeta}{\omega_n\Delta t}(1 - e^{-\zeta\omega_n\Delta t}\cos \omega_d\Delta t) - \frac{(1-2\zeta^2)e^{-\zeta\omega_n\Delta t}\sin \omega_d\Delta t}{\omega_d\Delta t}\right] \tag{6.61}$$

$$B_5 = -\frac{1}{2\omega_n^2}\left\{-\frac{4\zeta}{\omega_n\Delta t} - \left[\frac{2(1-4\zeta^2)}{\omega_n^2\Delta t^2} - \frac{2\zeta}{\omega_n\Delta t}\right](1 - e^{-\zeta\omega_n\Delta t}\cos \omega_d\Delta t)\right.$$

$$\left. + \left[\frac{1-2\zeta^2}{\omega_n\Delta t} + \frac{2\zeta(3-4\zeta^2)}{\omega_n^2\Delta t^2}\right]\frac{e^{-\zeta\omega_n\Delta t}\sin \omega_d\Delta t}{\sqrt{1-\zeta^2}}\right\} \tag{6.62}$$

$$B_6 = -\omega_n B_2 \tag{6.63}$$

$$B_7 = \frac{e^{-\zeta\omega_n\Delta t}}{\omega_n}\left(\cos \omega_d\Delta t - \frac{\zeta}{\sqrt{1-\zeta^2}}\sin \omega_d\Delta t\right) \tag{6.64}$$

$$B_8 = -\frac{B_2}{\omega_n} \tag{6.65}$$

$$B_9 = \frac{B_1-1}{\omega_n^3\Delta t} \tag{6.66}$$

$$B_{10} = \frac{1}{2\omega_n^2}\left\{\frac{2}{\omega_n\Delta t} - \left(\frac{1}{\omega_n\Delta t} + \frac{4\zeta}{\omega_n^2\Delta t^2}\right)(1 - e^{-\zeta\omega_n\Delta t}\cos \omega_d\Delta t)\right.$$

$$\left. - \left[\frac{2(1-2\zeta^2)}{\omega_n^2\Delta t^2} - \frac{\zeta}{\omega_n\Delta t}\right]\frac{e^{-\zeta\omega_n\Delta t}\sin \omega_d\Delta t}{\sqrt{1-\zeta^2}}\right\}. \tag{6.67}$$

This method has significant advantages over the numerical integration technique presented in the preceding section. First of all, computer running time is drastically reduced. The computation required to produce the relative deflection and velocity responses at each digitized point of the excitation consists of ten multiply–adds or approximately 210 μsec on the IBM 7094.

Another advantage is that both the real and the imaginary parts of the Fourier transform may be calculated from ξ and $\dot{\xi}$ when the damping ratio has been set to zero. This can be seen by a reexamination of Eq. (4.21) for $\zeta = 0$;

$$\xi(t) = \xi_0 \cos \omega_n t + \frac{\dot{\xi}_0}{\omega_n} \sin \omega_n t - \frac{1}{\omega_n} \int_0^t \ddot{x}(\tau) \sin \omega_n(t-\tau) dt. \qquad (6.68)$$

If $\xi_0 = \dot{\xi}_0 = 0$ and the trigonometric substitution

$$\sin(a-b) = \sin a \, \cos b - \cos a \, \sin b$$

is made, then

$$\omega_n \xi(t) = -\sin \omega_n t \int_0^t \ddot{x}(\tau) \cos \omega_n \tau d\tau + \cos \omega_n t \int_0^t \ddot{x}(\tau) \sin \omega_n \tau d\tau. \qquad (6.69)$$

Taking the derivative of Eq. (6.69) yields

$$\dot{\xi}(t) = -\cos \omega_n t \int_0^t \ddot{x}(\tau) \cos \omega_n \tau d\tau - \sin \omega_n t \int_0^t \ddot{x}(\tau) \sin \omega_n \tau d\tau. \qquad (6.70)$$

If $\xi(t)$ and $\dot{\xi}(t)$ are known for a particular time $t = T_D$, it is now possible to solve Eqs. (6.69) and (6.70) simultaneously to obtain the finite Fourier transform of the exciting function over the time range $t = 0$, T_D. To obtain the complete range Fourier transform, it is necessary to set T_D equal to or greater than the time at which $\ddot{x}(t)$ decays to zero. Therefore,

$$\int_0^{T_D} \ddot{x}(\tau) \cos \omega_n \tau d\tau = -\dot{\xi}(T_D) \cos \omega_n T_D - \omega_n \xi(T_D) \sin \omega_n T_D \qquad (6.71)$$

and

$$\int_0^{T_D} \ddot{x}(\tau) \sin \omega_n \tau d\tau = \omega_n \xi(T_D) \cos \omega_n T_D - \dot{\xi}(T_D) \sin \omega_n T_D. \qquad (6.72)$$

Or, in simplified notation,

$$Re\{F(\omega_n)\} = -\omega_n \xi(T_D) \cos \omega_n T_D - \dot{\xi}(T_D) \sin \omega_n T_D$$

$$Im\{F(\omega_n)\} = \omega_n \xi(T_D) \cos \omega_n T_D - \xi(T_D) \sin \omega_n T_D. \qquad (6.73)$$

A possible drawback of this method is the assumption that the parabolic approximation utilized is a good fit to the excitation data. If this assumption is to hold, a sampling frequency significantly greater than that specified by the sampling theorem is required.

The error introduced by this method can be seen by comparing the parabolic fits and the actual excitation as shown in Fig. 6.5, when the sampling frequency used approximates two samples per period of the highest data frequency.

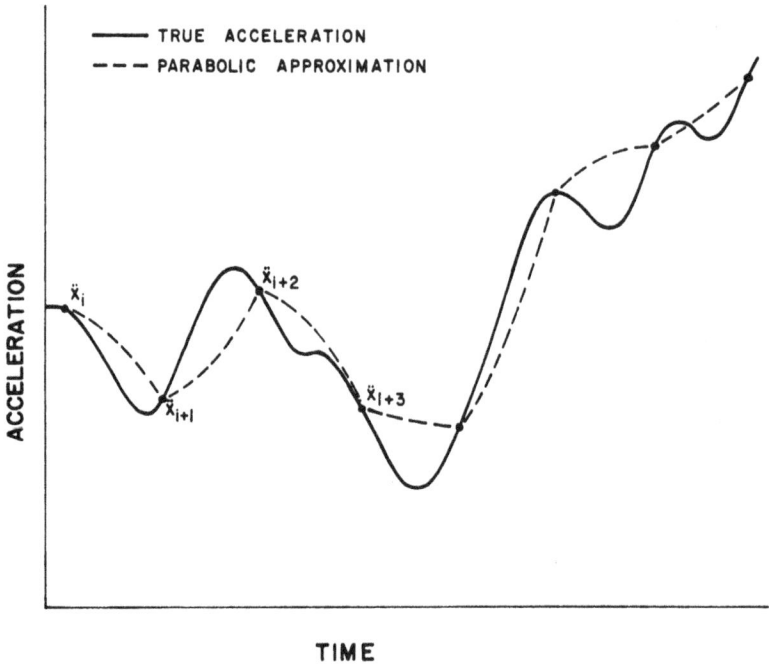

Fig. 6.5. Parabolic approximations to true acceleration for low relative sampling rate.

6.7 Response History Computation via Filtering

Undoubtedly, the fastest method for computing the response time history of a single degree-of-freedom system is by means of digital filters. A filter may be defined in general terms as a process which operates on a time history and changes the characteristics of that history in some specified manner. The filters to be discussed here are all of a linear nature and as such correspond to the linear systems described in Chapter 3.

There are several ways in which a linear filter may be defined. Probably the most basic definition is through the combination of the frequency response function and its inverse Fourier transform, the unit impulse response function. That is, if a filter is specified by its frequency response function $H(f)$, then its unit impulse response function is defined by

$$h(t) = \int_{-\infty}^{\infty} H(f)e^{j2\pi ft}df. \tag{6.74}$$

With $h(t)$ known, it is possible to generate the response of the physical system for any given input $x(t)$ by convolving it with $h(t)$;

$$y(t) = \int_0^t h(\tau)x(t-\tau)d\tau. \tag{6.75}$$

Nonrecursive Filters

In open-loop or nonrecursive digital filtering, this process is converted to the discrete form and automated. As an example, consider the base-excited, single degree-of-freedom system with absolute acceleration for both excitation and response. The discrete frequency response function is

$$H_\ell = H(\ell \Delta f) = \frac{1 + j2\zeta \ell \Delta f/f_n}{1 - (\ell \Delta f/f_n)^2 + j2\zeta \ell \Delta f/f_n}, \quad \ell = 0, 1, \ldots, M_1. \tag{6.76}$$

The inverse transform is computed from

$$h_i = h(i\Delta t) = \Delta f \sum_{\ell=0}^{M_1} \frac{1 + j2\zeta \ell \Delta f/f_n}{1 - (\ell \Delta f/f_n)^2 + j2\zeta \ell \Delta f/f_n} e^{j2\pi \ell \Delta f i \Delta t}$$

$$i = 0, 1, 2, \ldots, M_2, \tag{6.77}$$

where $M_1 \Delta f = f_c$, the Nyquist frequency. Note that the unit impulse response function is truncated at some finite point $M_2 \Delta t$. The filter weights calculated from Eq. (6.77) have the form

$$h_i = \frac{2\pi f_n i \Delta t}{\sqrt{1-\zeta^2}} e^{-\zeta 2\pi f_n i \Delta t} \sin\left(2\pi f_n i \Delta t \sqrt{1-\zeta^2} + \tan^{-1}\frac{\sqrt{1-\zeta^2}}{1/2-\zeta^2}\right),$$

$$i = 0, 1, 2, \ldots, M_2. \tag{6.78}$$

The response of the system may then be calculated from the discrete form of the convolution integral,

$$y_k = \Delta t \sum_{i=0}^{M_2} h_i x_{k-i}. \tag{6.79}$$

While this procedure is straightforward, it has a major drawback. For lightly damped systems, the unit impulse response decays slowly. This means that many filter weights must be used in order to maintain reasonable accuracy. In fact, it is quite likely that 50 to 100 weights must be used when the sampling frequency $1/\Delta t$ is less than $1/5$ the natural frequency of the system. The computer time required to perform such an analysis quickly becomes exorbitant. The filter may also become unstable because of a computer problem known as underflow, where the products of some of the smaller weights with the values of the excitation are so small as to be lost out of the arithmetic register during the summation process.

Recursive Filters

A type of filter which minimizes both of these problems is known as a recursive or feedback filter. This type of filter utilizes past values of the response as well as past and present values of the excitation in calculating the present response value. The form of this type of filter is

$$y_k = \sum_{i=0}^{M_1} a_i x_{k-i} + \sum_{\ell=1}^{M_2} b_\ell y_{k-\ell}. \tag{6.80}$$

Again, the basic procedure in utilizing such a filter requires the derivation of the filter weights a_i and b_ℓ. Equation (6.80) may be rewritten

$$y_k - \sum_{\ell=1}^{M_2} b_\ell y_{k-\ell} = \sum_{i=0}^{M_1} a_i x_{k-i}. \tag{6.81}$$

Taking Fourier transforms of both sides of the equation, we have

$$Y(f)\left[1 - \sum_{\ell=1}^{M_2} b_\ell e^{-j2\pi f\ell\Delta t}\right] = X(f)\sum_{i=0}^{M_1} a_i e^{-j2\pi f i\Delta t}. \tag{6.82}$$

The frequency response function of the filter is then

$$H(f) = \frac{Y(f)}{X(f)} = \frac{\displaystyle\sum_{i=0}^{M_1} a_i e^{-j2\pi f i\Delta t}}{1 - \displaystyle\sum_{\ell=1}^{M_2} b_\ell e^{-j2\pi f_\ell \Delta t}}. \tag{6.83}$$

As can be seen, both the numerator and denominator of the right-hand side of Eq. (6.83) are polynomials in $\exp[-j2\pi f\Delta t]$. In order to define the frequency response function, it is only necessary to determine the roots of these polynomials. A simplification of notation from digital control theory may be used at this point to make the polynomials more apparent;

$$z = e^{j2\pi f\Delta t}$$

$$H(f) = \frac{\displaystyle\sum_{i=0}^{M_1} a_i z^{-i}}{1 - \displaystyle\sum_{\ell=1}^{M_2} b_\ell z^{-\ell}}. \tag{6.84}$$

This procedure and the associated operations used in solving for the polynomial roots are usually termed z-transform theory. These roots are called either zeros or poles, depending on whether they are roots of the numerator or denominator. The important point to note about this form of the frequency response function is that the polynomial coefficients are exactly the filter weights required in order to perform the recursive filtering operation in the time domain. These coefficients are usually determined in one of two ways:

- By knowing the filter frequency response function and rewriting it in terms of polynomials in exp $[-j2\pi f \Delta t]$.
- By knowing the values of the zeros and poles of the filter frequency response function and expanding them to produce the required coefficients.

An example of each procedure will be shown as applied to different forms of the single degree-of-freedom system.

The first example is due to Lane [35] and utilizes the acceleration excitation–acceleration response version of the base-excited system. Again, the differential equation of this system is

$$m\ddot{y} + c\dot{y} + ky = c\dot{x} + kx, \tag{6.85}$$

with a frequency response function

$$H(f) = \frac{j2\zeta(f/f_n) + 1}{1 - (f/f_n)^2 + j2\zeta(f/f_n)}. \tag{6.86}$$

By making the substitutions

$$s = j2\pi f, \quad \omega_n = 2\pi f_n,$$

the filter transfer function in terms of the Laplace transform variable s is obtained;

$$H(s) = \frac{2\zeta\omega_n s + \omega_n^2}{s^2 + 2\zeta\omega_n s + \omega_n^2}. \tag{6.87}$$

The z-transform of $H(s)$ may then be computed from the relationship

$$H(z) = Z[H(s)] = \sum_{i=1}^{N} \frac{A(s_i)}{B'(s_i)} \frac{1}{1 - z^{-1}e^{-s_i \Delta t}}, \tag{6.88}$$

where $H(s) = A(s)/B(s)$, s_i is the ith pole of $H(s)$, and $B'(s)$ is the first derivative of $B(s)$.

Rewriting Eq. (6.87) in light of Eq. (6.88) produces the relationship

$$H(z) = d \left\{ \frac{-d - jg + c}{-jg[1 - z^{-1}e^{(-d-jg)\Delta t}]} + \frac{-d + jg + c}{jg[1 - z^{-1}e^{(-d+jg)\Delta t}]} \right\}, \qquad (6.89)$$

where

$$c = \frac{\omega_n}{2\zeta}$$

$$d = \zeta\omega_n$$

$$g = \omega_n \sqrt{1 - \zeta^2}.$$

By means of algebraic manipulations, Eq. (6.89) may be rewritten as

$$H(z) = \frac{2d + \left\{ 2de^{-d\Delta t} \left[\left(\frac{c-d}{g} \right) \sin g\Delta t - \cos g\Delta t \right] \right\} z^{-1}}{1 + (-2e^{d\Delta t} \cos g\Delta t)z^{-1} + (e^{-2d\Delta t})z^{-1}} \qquad (6.90)$$

or

$$H(z) = \frac{p_0 + p_1 z^{-1}}{1 - q_1 z^{-1} - q_2 z^{-2}}, \qquad (6.91)$$

where

$$p_0 = 2\zeta\omega_n = 4\pi\zeta f_n\Delta t \qquad (6.92)$$

$$p_1 = 4\zeta\omega_n\Delta t e^{-\zeta\omega_n\Delta t} \left[\frac{1 - 2\zeta^2}{2\zeta \sqrt{1 - \zeta^2}} \sin(\omega_n\Delta t\sqrt{1 - \zeta^2}) - \cos \omega_n\Delta t\sqrt{1 - \zeta^2} \right]$$

$$= 4\pi\zeta f_n\Delta t e^{-2\pi\zeta f_n\Delta t} \left[\frac{1 - 2\zeta^2}{2\zeta \sqrt{1 - \zeta^2}} \sin(2\pi f_n\Delta t\sqrt{1 - \zeta^2}) - \cos 2\pi f_n\Delta t\sqrt{1 - \zeta^2} \right]$$
$$\qquad (6.93)$$

$$q_1 = 2e^{-\zeta\omega_n\Delta t} \cos \omega_n\Delta t\sqrt{1 - \zeta^2}$$

$$= 2e^{-2\pi\zeta f_n\Delta t} \cos 2\pi f_n\Delta t\sqrt{1 - \zeta^2} \qquad (6.94)$$

$$q_2 = -e^{-2\zeta\omega_n\Delta t} = -e^{-4\pi\zeta f_n\Delta t}. \qquad (6.95)$$

A scale factor of Δt is usually included in the nonrecursive weights p_0 and p_1 to normalize them with respect to the sampling rate. The filter is applied as described in Eq. (6.96):

$$y_k = \sum_{i=0}^{1} -p_i x_{k-i} + \sum_{\ell=1}^{2} q_\ell y_{k-\ell}. \qquad (6.96)$$

As can be seen, the entire procedure requires only four multiplications and four additions in order to generate one response value. For reasonable record lengths $(1000 \leqslant N \leqslant 10{,}000)$, this filtering process requires less than 2 sec of IBM 7094 time per response history when input-output time is ignored.

The second example to be discussed is due to Otnes [36]. This filter utilizes the acceleration excitation-relative displacement response version of the base-excited system. The differential equation is

$$\ddot{\xi} + 2\zeta\omega_n\dot{\xi} + \omega_n^2\xi = \ddot{x}(t). \tag{6.97}$$

The required frequency response function relates ξ to $x(t)$;

$$H(f) = \frac{1}{(\omega_n^2 - \omega^2) + j(2\zeta\omega\omega_n)}. \tag{6.98}$$

Note that this function has only two poles and no zeros. Therefore, an appropriate form for the recursive filter is

$$\xi_k = a\ddot{x} + \sum_{i=1}^{2} h_i\xi_{k-i}. \tag{6.99}$$

To determine the values of the filter coefficients, if is necessary to express $H(f)$ in the form

$$H(f) = \frac{a}{1 - \sum_{i=1}^{2} b_i e^{-j2\pi f_i \Delta t}}. \tag{6.100}$$

This can best be done by determining the poles of Eq. (6.98), replacing them by their discrete forms, and then taking their products to produce the required polynomial. The poles of this system are complex and, in fact, are conjugates. If one of the poles is defined as Λ_1, then

$$\begin{aligned} \Lambda_1 &= \alpha + j\beta \\ \Lambda_2 &= \Lambda_1^* = \alpha - j\beta. \end{aligned} \tag{6.101}$$

The coefficients α and β specify the placement of the poles in the complex plane and are usually denoted as functions of the natural circular frequency ω_n and damping ratio ζ;

$$\alpha = -\zeta\Delta t\omega_n = -2\pi\zeta\Delta t f_n$$

$$j\beta = j\Delta t\sqrt{1 - \zeta^2}\omega_n = j2\pi\Delta t\sqrt{1 - \zeta^2}f_n, \tag{6.102}$$

where ω_n is the system undamped natural frequency in radians and f_n is the same frequency in hertzes. The two poles of the system are therefore

$$\Lambda_1 = j2\pi\Delta t\sqrt{1-\zeta^2}f_n - 2\pi\zeta\Delta tf_n$$

$$= 2\pi\Delta tf_n(j\sqrt{1-\zeta^2}-\zeta) \tag{6.103}$$

and

$$\Lambda_2 = \Lambda_1^* = -2\pi\Delta tf_n(j\sqrt{1-\zeta^2}+\zeta). \tag{6.104}$$

The denominator of the frequency response function may now be ₁ defined as a polynomial in exp $[-j2\pi\Delta t]$ as follows:

$$p(f) = [1 - e^{(\Lambda_1 - j2\pi f\Delta t)}]\,[1 - e^{(\Lambda_2 - j2\pi f\Delta t)}]$$

$$= [1 - e^{-j2\pi f\Delta t}\,e^{2\pi f_n\Delta t(j\sqrt{1-\zeta^2}-\zeta)}]\,[1 - e^{-j2\pi f\Delta t}\,e^{-2\pi f_n\Delta t(j\sqrt{1-\zeta^2}+\zeta)}]$$

$$= [1 - e^{-j2\pi f\Delta t}\,e^{-2\pi\zeta f\Delta t}\,(e^{-j2\pi f_n\Delta t\sqrt{1-\zeta^2}} + e^{j2\pi f_n\Delta t\sqrt{1-\zeta^2}})$$

$$+ e^{-4j\pi f\Delta t}\,e^{-4\pi\zeta f_n\Delta t}]$$

$$= 1 - e^{-j2\pi\zeta f\Delta t}\,[2\cos(2\pi f_n\Delta t\sqrt{1-\zeta^2})\,e^{-2\pi\zeta f_n\Delta t}]$$

$$+ e^{-j4\pi\zeta f\Delta t}\,(e^{-4\pi\zeta f_n\Delta t}). \tag{6.105}$$

The coefficients of the feedback terms in the digital filter are simply the polynomial coefficients

$$b_1 = 2e^{-2\pi\zeta f_n\Delta t}\cos(2\pi f_n\Delta t\sqrt{1-\zeta^2}) \tag{6.106}$$

$$b_2 = -e^{-4\pi\zeta f_n\Delta t}. \tag{6.107}$$

The coefficient of the nonrecursive term is simply a multiplicative constant and may be determined by noting that at $f=0$ the modulus squared of the frequency response function must equal $1/(16\pi^4 f_n^4)$. Therefore, from Eq. (6.102)

$$\frac{1}{16\pi^4 f_n^4} = \frac{a^2}{(1-b_1-b_2)^2} \tag{6.108}$$

or

$$a = \frac{1-b_1-b_2}{4\pi^2 f_n^2}. \tag{6.109}$$

This version of the single degree-of-freedom filter is slightly faster to compute than the preceding example in that only three multiplications and three additions are required per response value. It should be noted, however, that it is not possible to obtain the absolute acceleration response with this filter. Instead, the equivalent static acceleration must be used as the response parameter.

6.8 Peak Detection Methods

One of the major problems associated with the determination of a shock spectrum is to define accurately the peak response from the available samples of the response time history. As will be noted in the next section, selecting the observed maximum as an estimate of the true peak leads to bias errors, especially at the higher frequencies. Aside from the brute-force solution of increasing the sampling frequency of the response history to the point where the bias error is reduced to an acceptable level, the only other procedure for bias error minimization available requires the use of an interpolation formula.

Interpolation may be thought of as the determination of some well-defined analytic function which approximates the sampled data to some desired degree of accuracy. This analytic function may then be used to determine required values which have not originally been provided.

A reasonable procedure to follow when defining the type of interpolation function is to examine the process used to generate the sampled time history. In the case of the response of a single degree-of-freedom system, it is obvious that a sinusoid would be an appropriate interpolator. Unfortunately, the use of trigonometric functions for peak detection presents problems.

Determining the peak response consists of two separate and distinct operations. The first of these is to locate in time the relative maxima and minima of the response history. The second operation is the evaluation of these maxima and minima.

The usual procedure followed is to evaluate the coefficients of the interpolating function by utilizing a set of the sampled response values and solving the resultant linear equations. When this approach is attempted with trigonometric functions, the equations generated are transcendental in nature and cannot be solved explicitly for the required coefficients.

Instead of using a sinusoid as the interpolating function, it is usually sufficient to use a series expansion for the sine or cosine. This immediately gives rise to polynomial interpolation. Postulating an interpolating polynomial yields

$$p(t) = \sum_{i=0}^{M} a_i t^i. \tag{6.112}$$

Then $M+1$ samples of the response history are required to uniquely define the polynomial coefficients. The coefficients may be obtained by simultaneously solving the set of equations

$$r_k = \sum_{i=0}^{M} a_i t_k^i, \qquad k = 1, 2, 3, \ldots, M+1, \tag{6.113}$$

where r_k is the sampled response for the a_i's. It is now possible to utilize this polynomial to both detect and evaluate all relative maxima and minima of the response history. By differentiating the interpolating polynomial and setting the derivative to zero, the times of relative extrema can be detected. As an example, consider the decaying sinusoid

$$y(t) = Ae^{-\alpha t} \sin (2\pi ft + \phi). \qquad (6.114)$$

For $A = 27$, $\alpha = 0.25$, $f = 1$ Hz, and $\phi = \pi/6$, the following sampled time history is obtained:

t	y
0	13.500
0.125	25.272
0.25	21.924
0.375	6.372
0.5	-11.907
0.625	-22.275
0.75	-19.359

The true absolute maximum will occur at $t = 0.167$ and will have a value of 26.433.

Utilizing the first three samples of the time history allows the derivation of a second degree polynomial. This polynomial has the form

$$\hat{y} = -483.84t^2 + 154.656t + 13.5. \qquad (6.115)$$

By taking the first derivative and setting it to zero,

$$-96.768t_E + 154.656 = 0 \qquad (6.116)$$

or

$$t_E = 0.1597.$$

The value of the relative maximum may now be determined by solving the interpolating polynomial at $t = 0.1597$.

$$\hat{y}_E = -483.84 \, (0.1597)^2 + 154.656 \, (1597) + 13.5 \qquad (6.117)$$

or

$$\hat{y}_E = 25.861.$$

It is possible to obtain another estimate of the same extremum simply by utilizing the three samples starting with the second value and performing the same procedure.

For this particular example, an error reduction of 40 percent over the simple selection of the observed maximum was obtained. Usual error reduction will be in the neighborhood of 20 to 30 percent.

The degree of the interpolation polynomial is effectively limited to five because of the need to solve explicitly the first derivative of the interpolator, and it is not possible to produce a general closed-form solution for equations greater than fourth degree.

A basic procedure for implementing this interpolator is what is termed a moving arc. The polynomial is applied to a set of samples, and any extreme points in the range of the samples are evaluated. Then the polynomial is shifted one sample and again evaluated. This procedure is continued until all possible sets of samples have been analyzed. As each extreme point is determined, it is compared with the previous point and only the largest value is retained.

For equally spaced samples and quadratic interpolation, the explicit formulas are as follows.

Given three samples of the response history r_0, r_1, and r_2 taken at times t_0, $t_0 + \Delta t$ and $t_0 + 2\Delta t$; then

$$a_0 = r_0 - \frac{t_0(r_1 - r_0)}{\Delta t} + \frac{(t_0 + \Delta t)(r_2 - 2r_1 + r_0)}{(t_0 + 2\Delta t)(t_0 + \Delta t) + 2t_0 + 5\Delta t} \tag{6.118}$$

$$a_1 = \frac{r_1 - r_0}{\Delta t} - \frac{(r_2 - 2r_1)(2t_0 + \Delta t)}{t_0[(t_0 + 2\Delta t)(t_0 + \Delta t) + 2t_0 + 5\Delta t]} \tag{6.119}$$

$$a_2 = \frac{r_2 - 2r_1 + r_0}{t_0[(t_0 + 2\Delta t)(t_0 + \Delta t) + 2t_0 + 5\Delta t]} \tag{6.120}$$

$$t_E = -\frac{a_1}{2a_2} \tag{6.121}$$

$$\hat{y}_E = a_0 + a_1 t_E + a_2 t_E^2. \tag{6.122}$$

Another procedure which may be used consists of somehow detecting the intervals containing the extreme points and then interpolating for enough additional samples in these intervals to guarantee the required accuracy. Lane [35] suggests the use of an integrating filter to detect the interval containing the peak response and then the use of a seventh-degree polynomial interpolator.

If the data are properly band-limited, it would appear that the so-called $(\sin x)/x$ function is a more appropriate interpolating function. It may be shown [27] that if a time history has no frequency components above $1/2\Delta t$, then the continuous time history may be obtained from the sampled values by

$$y(t) = \sum_{k=-\infty}^{\infty} y(k\Delta t) \left\{ \frac{\sin\left[(\pi/\Delta t)(t - k\Delta t)\right]}{(\pi/\Delta t)(t - k\Delta t)} \right\}. \tag{6.123}$$

Note that the summation limits are infinite. In the normal situation, these limits cover the available record length, but it is possible to truncate the series at approximately 20 terms without appreciable error.

6.9   Error Analysis

The error involved in performing a digital shock spectrum analysis may be categorized as

1. Error due to sampled excitation,
2. Error inherent in solution technique, or
3. Error due to sampled response.

The first of these classifications is basically concerned with the aliasing problem. Since anti-aliasing analog filters can have undesirable characteristics when the filter cutoff frequency is near the frequencies of interest, it is suggested that, if possible, the cutoff frequency be set to at least twice the highest frequency of interest. The excitation should then be sampled at two times this highest frequency to avoid aliasing. Digital low-pass filters may then be used on the digitized data to perform the band-limiting operation, and the data may be decimated to a more reasonable sampling frequency.

The second category is concerned with the errors in the technique used to calculate the response histories. Two different problem areas are involved here. The first of these is simply the error due to sampling the system impulse response function instead of using the continuous function. This is usually unimportant by comparison with the other errors. The other error source occurs in both the numerical integration and the recursion techniques. In each case, the assumption is made that the excitation may be well approximated by a series of straight-line segments. The effect of this assumption is to bias the response history, usually in a downward direction. This bias effect becomes more pronounced as the excitation sampling rate is reduced and the straight lines do a poorer job of fitting the data. This error is more noticeable at the low frequencies where the response is highly velocity-dependent, but the bias will occur at all frequencies.

Procedures for minimizing this type of error consist of

1. Increasing the sampling frequency to the point where the line segments do a good job of fitting the data (usually a minimum of 10 samples/cycle of the highest data frequency), or
2. Using a nonlinear fit to the sampled data.

Intuitively, it would appear that a sine approximation to the sampled data would be best, but the use of a transcendental function increases the complexity of the procedure considerably. A compromise between ease of programming and error reduction would be to use a quadratic to only fit the data. This would introduce a second difference term ($\Delta^2 x$) into the recursion formula and would cause a similar change in the numerical integration procedure.

The third error category is concerned with the difficulty in determining the true response peak from the sampled response time history. The simplest peak-detection technique is to search the response history samples for the maximum value and assume that this maximum is the

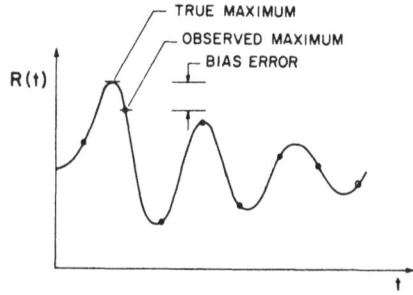

Fig. 6.6. Sampled response bias error.

true peak. As can be seen from Fig. 6.6, this inevitably leads to a bias error.

In fact, as has been pointed out by Lane [35], the true peak response may not even be in the vicinity of the observed peak. The methods for reducing this type of error consist of either increasing the sampling rate of the response or utilizing an interpolation formula. These procedures were discussed in Section 6.8.

Error estimates for interpolation procedures vary drastically with the interpolating function, so no discussion of this subject will be presented. However, it is possible to provide estimates of the bias error due to selecting the peak sampled response as an estimate of the true peak.

If one considers the response to be a sinusoid, which it is for zero damping, then it is possible to bound the error incurred by accepting the observed maximum as the true peak. The greatest error occurs when the true peak lies halfway between two samples of the response. This can be seen from Fig. 6.7.

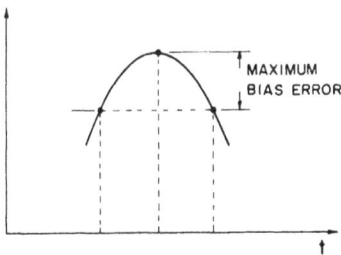

Fig. 6.7.   Maximum bias error.

Define M as the ratio of the sampling frequency to the response frequency;

$$M = \frac{f_s}{f_r}.$$

Then the maximum percentage error is

$$e(\%) = 100\left(1 - \cos\frac{\pi}{M}\right).$$

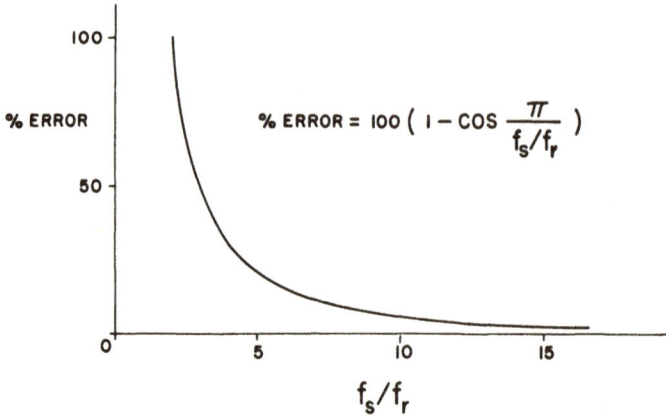

Fig. 6.8.   Maximum percentage error.

A plot of this error bound appears in Fig. 6.8. This bound tends to be quite conservative. For example, at 2 samples/cycle, the error bound is 100 percent.

A more reasonable estimate of the error may be obtained from a probabilistic approach. Again, assume that the response curve is a sinusoid,

$$x(t) = A \sin (2\pi f_r t + \phi), \tag{6.125}$$

where ϕ is the phase angle (assumed to be random) and A is the required peak amplitude. The observed sequence is x_i, where

$$x_i = x(i\Delta t) = A \sin (2\pi f_r i\Delta t + \phi). \tag{6.126}$$

The expected value of the observed peak $\hat{A}$ will be

$$
\begin{aligned}
\mathrm{E}[\hat{A}] &= \frac{A}{\Delta t} \int_{-\Delta t/2}^{\Delta t/2} \cos 2\pi f_r t \, dt \\
&= \frac{1}{\Delta t} \left[\frac{A}{2\pi f_r} \sin 2\pi f_r t \right]_{-\Delta t/2}^{\Delta t/2} \\
&= A \frac{\sin \pi f_r \Delta t}{\pi f_r \Delta t} .
\end{aligned}
\tag{6.127}
$$

In other words, the bias is what is commonly termed a $(\sin x)/x$ curve as shown in Fig. 6.9. The percentage expected error is

$$e(\%) = 100 \left(1 - \frac{\sin \pi f_r \Delta t}{\pi f_r \Delta t} \right), \tag{6.128}$$

with the maximum of approximately 36 percent occurring at half the sampling frequency.

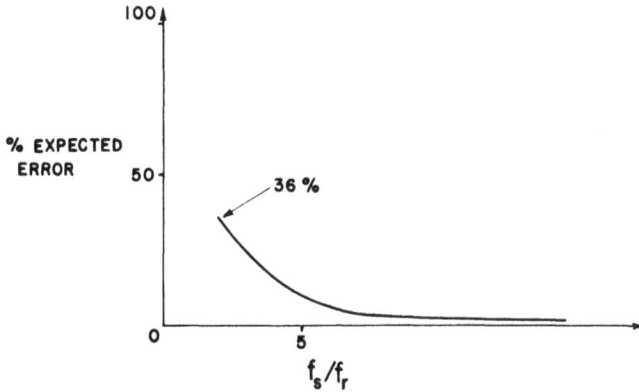

Fig. 6.9. Expected percentage error.

The 5-percent *expected* error point occurs approximately at a sampling frequency equal to six times the resonant frequency. This sampling requirement is considerably less stringent than the 10 sample/cycle requirement imposed by the error bound described earlier for a maximum error of 5 percent.

Chapter 7

MISCELLANEOUS TECHNIQUES

The majority of the preceding text has been devoted to describing two specific methods for analyzing shock data: the Fourier and shock spectral analyses. The reason for this is that these two methods are by far the most commonly employed techniques for reducing shock data. In this chapter, other techniques that have been used are described.

7.1 Nonspectral Techniques

The process of data reduction is that of condensing a quantity of data until only the important properties remain, or, in some cases, rearranging the information contained in the data so that the important properties are more apparent. The key to the type of data reduction that should be employed is the word "important" in the above sentence. This is a subjective measurement. For engineering work, important properties should be interpreted as meaning those properties that can be profitably used to arrive at a solution to the engineering problem under investigation. Since the goal of data reduction is to assist in the solution of some specific problem, the data reduction technique that should be employed is the one that is most profitable (where profitability is based on a criterion or set of criteria, such as minimum cost, minimum time to solution, etc.). Sometimes when solutions cannot be obtained from a single form of data reduction, a second type will produce enough additional information to solve the problem. Nonspectral analysis techniques are those analysis techniques whose results are not functions of frequency. In this section five different types of nonspectral analyses are discussed. These are single-number, velocity-change, time-function decomposition, waveform integration, and phase-plane analysis techniques.

Single-Number Analyses

The simplest data reduction that can be used is to condense a complicated time history to a single number. For example, consider the acceleration time history of a mechanical shock shown in Fig. 7.1.

AMPLITUDE

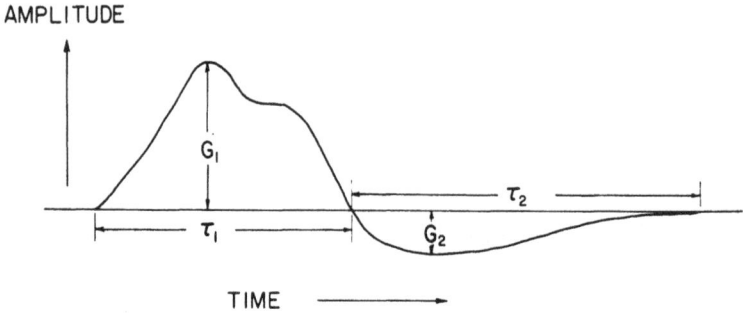

Fig. 7.1.   Shock acceleration time history.

Examples of single amplitude values that could be used are the highest positive peak G_1, the highest negative peak G_2, and the maximum peak-to-peak $(G_1 + G_2)$ acceleration. Examples of single duration measurements that can be used are the time to the first zero crossing (τ_1), the time to the second zero crossing (τ_2), and in this case the total time duration of the pulse $(\tau_1 + \tau_2)$.

Whether any of these simple measurements are sufficient to solve a problem or are of absolutely no value depends on the details of the specific problem. Clearly, when the pulse shape is fixed and only its amplitude and/or duration varies from shock to shock, any one of the above amplitude and/or duration measurements can be used to compare the shocks. Perhaps the complete Fourier spectrum is defined by the first shock; then, the ratio of the amplitude of successive shocks can be used to scale the amplitude of the Fourier spectrum. Similarly, the duration ratio can be used to scale the frequency.

These single properties can occasionally be used where sufficient empirical correlation exists between them and the performance of the system. However, there is a tacit requirement that these shocks all be of approximately the same wave shape. The response of most mechanical systems to two 10-g peak-to-peak transients will be drastically different if one is basically a single pulse with a 1-sec duration, and the other is a damped oscillatory type of pulse with a 1-msec overall duration.

Simple amplitude and duration measurements can sometimes be combined with some knowledge of the physical system to estimate responses. For example, if it is known that the durations of the pulses $(T_1$ and T_2 in Fig. 7.1) are quite long compared to the period of the first resonant frequency of the physical system, then the peak response of the system will be essentially equal to the peak of the input. The system sees the input almost as though it were a static input, and the output response will look very much like the input.

If the shock is primarily a single pulse of one polarity and the duration is quite short compared to the period of the highest frequency passed by the physical system, these simple amplitude and duration parameters can be used to predict the response of the system. In this case, the time history on the output of the system will be very similar to the weighting function of the physical system. The output response can be estimated by multiplying the weighting function of the system by the area under the input shock pulse.

Velocity Change Analyses

This leads to another relatively simple form of data reduction. That is a measure of the velocity change caused by the shock. The velocity change is a measure of the energy imparted by the shock. The physical system is assumed to be an undamped simple mechanical oscillator; it will oscillate continuously after the shock is removed. As the mass of the system oscillates, there is an interchange of energy from kinetic to potential energy. At the instant when the mass is at the position of rest (zero relative displacement in the mechanical oscillator), the velocity is at a maximum and all the energy is kinetic energy. As the mass passes this position, some of the kinetic energy is transferred to potential energy. At the point where the relative velocity of the mass reaches zero, the mass attains its maximum deflection and all of the energy is potential energy.

The energy imparted by the shock if the system is initially at rest is

$$E = (\tfrac{1}{2})\ m\Delta V^2, \tag{7.1}$$

where

E = the energy

m = the mass of the physical system

ΔV = the change in velocity caused by the shock.

Since conservation of energy is assumed, the sum of the instantaneous kinetic and potential energies will equal the total energy imparted;

$$K.\,E.\ (t) + P.\ E.\ (t) = E \tag{7.2}$$

$$(\tfrac{1}{2})\ mv\,(t) + (\tfrac{1}{2})\ kx^2\,(t) = (\tfrac{1}{2})\ m\Delta V^2,$$

where

$K.\,E.\ (t)$ = the instantaneous kinetic energy

$P.\,E.\ (t)$ = the instantaneous potential energy

k = the spring constant of the physical system

The maximum acceleration response can be found as follows:

$$a_{max} = \frac{F_{max}}{m}$$

$$= \frac{kx_{max}}{m}. \tag{7.3}$$

From Eq. (7.2), the maximum deflection can be found when the potential energy is a maximum and the kinetic energy is zero;

$$(\tfrac{1}{2})\, kx^2_{max} = (\tfrac{1}{2})\, m\Delta V^2$$

$$x_{max} = \Delta V \sqrt{\frac{m}{k}}. \tag{7.4}$$

Combining this result and Eq. (7.3) yields the maximum acceleration;

$$a_{max} = \Delta V \sqrt{\frac{k}{m}} = \Delta V 2\pi f_n,$$

where

> f_n = the undamped natural frequency of the physical system.
>
> $m = A_1 - A$.

Thus to estimate the maximum acceleration response of a structure, the change in velocity is calculated and this is multiplied by 2π times the natural frequency of the mechanical system.

Since all maximax shock spectra do not increase linearly with frequency, it is easy to see that this velocity change analysis does not apply to all shocks. This technique should only be used when the period of the natural frequency of the physical system is long compared to the duration of the shock. Basically, the method assumes a step change in velocity, and as the frequency of the physical system increases relative to the duration of the shock, the actual details of the waveform assume more importance.

Time Function Decomposition

Another approach employed to analyze the response of physical systems to shocks consists of decomposing the time history into one or several simple time functions. This process is fairly well defined for a clear-cut transient such as the N-wave associated with sonic-boom measurements. In Fig. 7.2a an N-wave is replotted from Ref. 37. This shock time history can be synthesized by combining four simple functions. These are a positive step function of amplitude A at time zero, a linear segment starting at time zero and having slope $(A_2 - A_1)/T$, a positive step function of amplitude A_2 at time T, and a linear segment starting at time T and having a slope that is the negative of the pre-

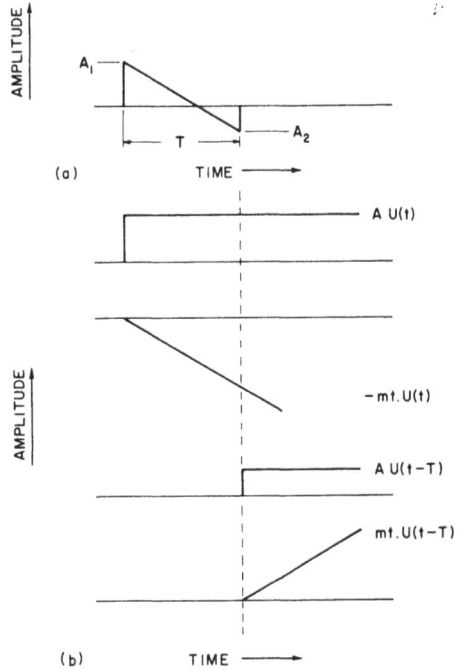

Fig. 7.2. Decomposition of a sonic boom N-wave, where m = (A₁ − A₂)/T.

vious linear segment. This synthesized time history is described in Eq. (7.6) and Fig. 7.2b;

$$x(t) \approx u(t)A_1 + \frac{A_2 - A_1}{T} t + u(t-T)\left(A_2 + \frac{A_1 - A_2}{T}\right), \qquad (7.6)$$

where

$$u(t) = \text{the unit step function.}$$

This synthesized time history is then used as a forcing function, or input, to the equations used to describe the response of the system.

The accuracy of this technique is primarily dependent on two factors. The first factor is the complexity of the shock time history. The more complicated the time history, the harder it is to fit simple mathematical functions to it, and obviously, the less accurate the model of the input will be. Figure 7.3 compares the N-wave of Fig. 7.2 to two other N-wave measurements utilizing the same aircraft [37]. For all three of these time histories one might use the synthesized input of Eq. (7.6). While this leads to a reasonably accurate description of the time history in Fig. 7.3a, it will have insufficient high-frequency content for the time history of Fig. 7.3b, and excessive high-frequency content for the time history in Fig. 7.3c.

(a)

(b)

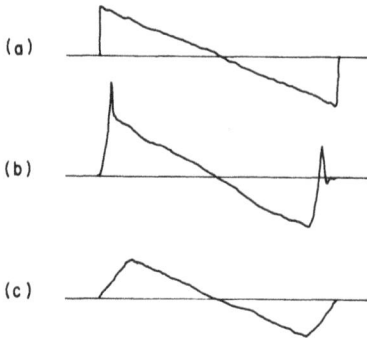

Fig. 7.3. Typical sonar boom time histories.

(c)

The second factor is the relation of the shock duration to the lowest resonant frequency in the physical system. The shorter the duration of the shock relative to the period of this first resonant frequency, the less accurate this approach becomes. Factors such as the exact phasing between simple components become quite critical. However, this method can lead to more accurate response calculations than the simple techniques previously described in this chapter.

Waveform Integration

Another relatively simple nonspectral data reduction technique that is used [38] is integration and double integration of the time history. The basic idea of this technique is to rearrange the information contained in the original data so that the important properties are more clearly visible. Both integration and double integration are low-pass filtering operations, so that the low frequency information is emphasized and the high frequency information deemphasized. Integration is a $1/f$ low-pass operation, and double integration is a $1/f^2$ low-pass operation. These particular forms of low-pass filtering are such that the filtered signals have physical significance. Starting with an acceleration time history, integration yields a velocity time history and double integration yields a displacement time history.

Figure 7.4 is reproduced from Ref. 38. In this figure an acceleration time history, its integral, and its double integral are compared. The velocity time history proved to be the most valuable in this case, as the increasing amplitude oscillations in the velocity record revealed the presence of two coupled modes, very closely spaced in frequency, in the structure.

Phase Plane Analyses

Given the time history of the input and a knowledge of the system, the output of the system can be calculated by convolving the input and the weighting function of the system as described in Chapter 3.

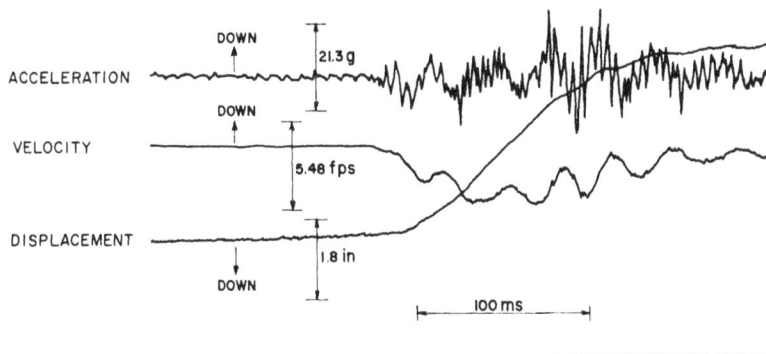

Fig. 7.4.   Integration and double integration of an acceleration time history.

One graphical method of evaluating the convolution integral was illustrated in Chapter 2. Another method for graphically computing the time history response of second order mechanical systems is the phase-plane method [39, 40].

This method consists of approximating the input time history by a series of contiguous step functions, projecting this approximation onto the phase plane to form a phase-plane trajectory, and then projecting back from the phase-plane trajectory to form the output time history.

A phase plane is a plot of the displacement of a system as a function of its velocity divided by the undamped resonant frequency of the system. Figure 7.5 is the phase-plane plot of an undamped second order system responding to a step in base displacement of amplitude x_0. Assuming that the system has zero initial conditions, the step in base displacement causes the locus of the system response to trace a circle of radius equal to the amount of the step around the point $(x = x_0, V/(2\pi f_n) = 0)$. The time required to make one complete revolution is $\tau_0 = 1/f_n$. The output, the absolute displacement of the mass of the seismic system $[y(t)]$, is found by projecting the moving point on the circle horizontally as a function of time. This yields the sine function shown in Fig. 7.5.

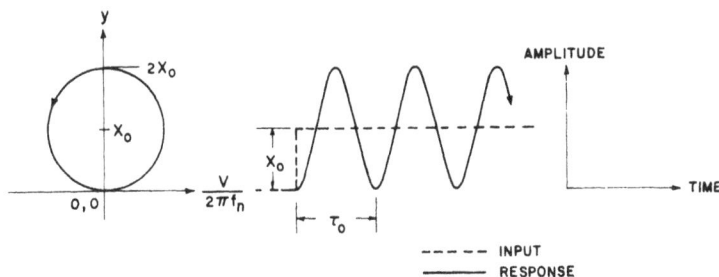

Fig. 7.5.   Phase-plane trajectory (undamped second order system).

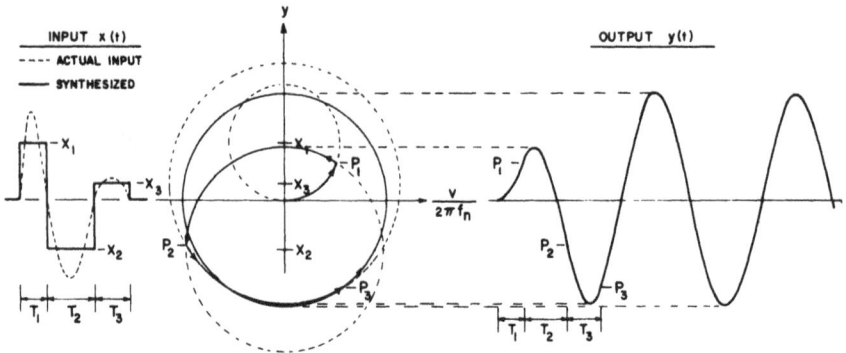

Fig. 7.6. Phase-plane plot; four-step function approximation of the input.

The absolute displacement response of the mass for a base displacement input is illustrated in Fig. 7.6. In this case, a time history is shown approximated by four step functions, at $t = 0$, T_1, T_2, and T_3. Assuming that the system is initially at rest, the first step function causes the phase-plane trajectory to rotate in a circle about $(x_1, 0)$ with radius x_1. In time T_1, it has rotated through an angle of $2\pi f_n T_1$. This is point P_1. At that time, the second step function causes the trajectory to rotate in a circle about the point $(x_2, 0)$. The radius is determined by the distance from $(x_2, 0)$ and P_1. At time $(T_1 + T_2)$, the trajectory has moved about this new circle an angle of $2\pi f_n T_2$ to point P_2. At that time, the third step function causes the trajectory to move in a circle about the point $(x_3, 0)$ with a radius equal to the distance between that point and P_2. The trajectory moves along this new circle through an angle of $2\pi f_n T_3$ to point P_3. At time $(T_1 + T_2 + T_3)$, the fourth step function occurs; this is the final one and is about zero. At that time, the trajectory moves onto a circle centered about $(0, 0)$ with a radius equal to the distance from 0 to P_3. It will continue to rotate on that circle indefinitely. The solid lines show the complete trajectory. The time history of the displacement response is obtained by projecting the amplitude y of the trajectory horizontally as a function of time. Likewise the velocity, scaled by $2\pi f_n$, can be obtained by projecting the time history of the trajectory vertically.

This technique and a modification of it called the phase-plane delta method can also be used to calculate graphically the response of nonlinear systems. While the accuracy obtainable with these methods does not compare with that of more complex analytical procedures, it is a very convenient method of manual analysis.

7.2 Analysis of Random Transients

Up to this point, all of the analysis methods have assumed that a single measurement is adequate to describe the shock. In this sense, the shocks considered are deterministic. The equation describing this shock applies to all other measurements of the same shock. In fact, all shock processes in nature are nonstationary random transients. An electronic oscillator does not oscillate at the same frequency and amplitude for all time. However, there are an enormous number of cases where the assumption of stationarity and even that of a deterministic form are completely justified from a practical point of view — the above oscillator, for instance.

Similarly, there are cases where the only justifiable form is that of a nonstationary random transient process. An example of this is the shock imparted to an aircraft when landing. The velocity of the aircraft, its attitude, the runway surface conditions, the ground winds, the weight of the aircraft at landing, and several other factors combine to determine each individual shock. Since all of these factors are variables, it is only reasonable to expect that the shock time histories from a number of landings would vary widely even if the measurements were restricted to a single aircraft and a single measuring location on that aircraft. Analyzing these data by any of the previously described methods results in wide variations of the data reduction results from landing to landing. This is highly undesirable since it greatly complicates, and in some cases, may even negate the application of the data reduction results to the solution of a problem.

A mathematically correct way to approach the problem is to collect a large number of records of the shock time histories and then perform ensemble averaging on the data. Ensemble averaging consists of sampling each record at the same instant of time from a starting time in each record (for example, the instant of touchdown could be used for $t = 0$ in the above case) and computing the statistical properties at that time. The time histories should be sampled at some sufficiently close separation, and the ensemble averaging should be performed at each time sample to obtain the statistical properties. Then, the joint statistical properties between values of this single process at different times should be calculated.

The exact statistical properties that must be measured for practical applications depend on two factors: the application of the results and the composition of the data. For shock problems, the applications have generally been those of determining the energy in the response of a structure or of determining the maximum response value caused by the shock. When the data have a Gaussian distribution, only the mean and variance need be computed completely to describe the first order (at a single instance of time) properties. (However, this is a unique situation.)

Time-Varying Mean Square Value

The time-varying mean square value of the response can be used as a measure of the damage potential of a shock, since the mean square value is related to the energy. If a linear, constant-parameter system is assumed, the response can be calculated from the input by the convolution integral

$$y(t) = \int_0^t x(\tau)h(t-\tau)d\tau. \tag{7.7}$$

The squared value of the response is

$$\begin{aligned}
y^2(t) &= \int_0^t x(\tau)h(t-\tau)\,d\tau \int_0^t x(u)h(t-u)\,du \\
&= \int_0^t \int_0^t x(\tau)x(u)h(t-\tau)h(t-u)\,d\tau\,du.
\end{aligned} \tag{7.8}$$

The mean square value is computed by taking the mathematical expectation of the square of the response value;

$$\begin{aligned}
\Psi_y^2(t) &= \mathbf{E}\left[\int_0^t \int_0^t x(\tau)x(u)h(t-\tau)h(t-u)\,d\tau\,du \right] \\
&= \int_0^t \int_0^t R_x(\tau,\,u)h(t-\tau)h(t-u)\,d\tau\,du,
\end{aligned} \tag{7.9}$$

where

$R_x(\tau,\,u) = \mathbf{E}[x(\tau)x(u)] = $ the nonstationary autocorrelation function.

Thus, the time-varying, mean square value of the response of a linear system can be calculated from an arbitrary nonstationary input if the weighting function of the system and the nonstationary autocorrelation function of the input are known. Thus, the data reduction problem becomes one of determining the autocorrelation function of the input. This is a rather complicated data reduction technique. To compute the nonstationary autocorrelation function, the value of each record at time t_1 and t_2 must be measured. These values are multiplied to form a product for each record. Then an ensemble average of the products must be computed. This average is computed by summing all these products and dividing by the number of products. This is shown in Eq. (7.10) and Fig. 7.7:

$$R_x(t_1,\,t_2) = \lim_{N \to \infty} \frac{1}{N} \sum_{i=1}^{N} x_i(t_1)\,x_i(t_2), \tag{7.10}$$

where

$i = $ the record number.

In practice, the limiting operation is dropped in Eq. (7.10), so that an estimate, rather than the true value of the autocorrelation, is found. The value of N must be large if the true value is to have a high probability of being reasonably close to the estimate. (Methods of

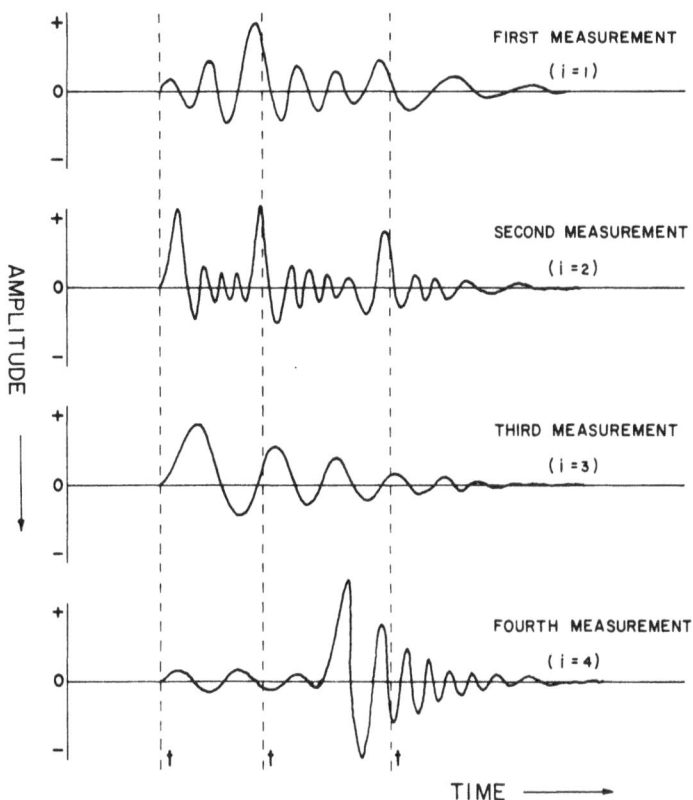

Fig. 7.7.   An ensemble of shock time histories.

determining the variance of nonstationary, mean square value estimates are discussed in Ref. 25 pp. 346, 347.) The data reduction operation
consists of defining a surface such as that shown in Fig. 7.8. Generally,
the surface will start at t_1 and $t_2 = 0$ and stop at t_1 and $t_2 = T$ (where
T = the duration of the shock). One model of a specific type of nonstationarity has been used to fit certain nonstationary data. This model
is composed of the product of two separable signals, one integrable and
one an arbitrary random signal. (In practice, the first signal is usually
assumed deterministic and the second signal a stationary, ergodic,
random signal.) The model is

$$x(t) = A(t) \cdot B(t),$$

where

$$A(t) = \text{the integrable signal}$$
$$B(t) = \text{the random signal}.$$

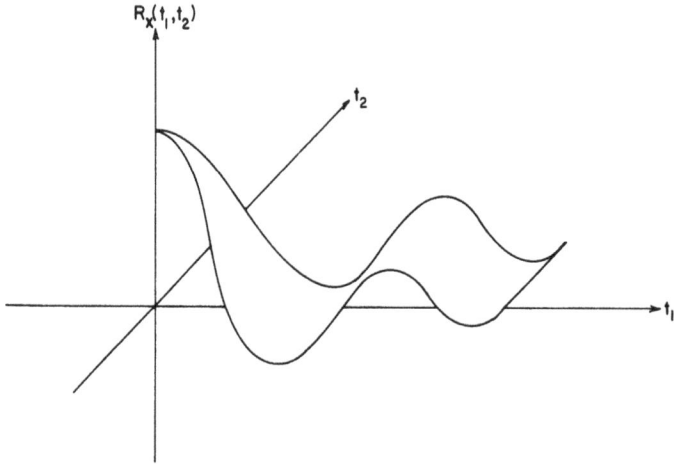

Fig. 7.8. Plot of a nonstationary autocorrelation function.

This simplifies the nonstationary autocorrelation function;

$$R_x(t_1, t_2) = \mathbf{E}[x(t_1)x(t_2)] = (A(t_1)A(t_2))\mathbf{E}[B(t_1)B(t_2)]$$

$$= A(t_1)A(t_2)R_B(t_1, t_2)$$

$$= A(t_1)A(t_2)R_B(\tau), \text{ when } B(t) \text{ is assumed stationary and er-godic } (\tau \text{ is the time separation, } (t_2-t_1)). \quad (7.11)$$

The time-varying, mean square response of a simple second order mechanical oscillator has been calculated for inputs with the above type of separable nonstationarity. See Refs. 41 and 42. These references examine the response of the mechanical oscillator to nonstationary signals that are the product of a deterministic function and a stationary random signal having a Gaussian distribution. The deterministic functions considered were
- A step function
- A boxcar (rectangular) function
- A decaying exponential function.

The random signals considered were
- White noise
- Bandpass filtered noise characterized by an exponential cosine autocorrelation function.

Numerous graphs are presented to depict the variation in the time-varying, mean square response with
- The resonant frequency of the mechanical system
- The critical damping ratio of the mechanical system
- The center frequency of the random signal
- The bandwidth of the random signal.

These references also demonstrated that, for correlated noise, the time-varying, mean square response can exceed its stationary response.

Generalized Spectral Density Function

The time-varying, mean square value can also be calculated from frequency domain information. The Fourier transform of the output of a simple linear system is the product of the Fourier transform of the input and the frequency response function of the system

$$Y_i(f) = H(f)X_i(f),$$

where

$Y_i(f)$ = the Fourier transform of the system response for the ith shock

$H(f)$ = the frequency response function of the system

$X_i(f)$ = the Fourier transform of the ith shock.

The time history of the response to the ith shock is the inverse transform of the Fourier spectrum of that response

$$y_i(t) = F^{-1}[Y_i(f)]$$

and

$$y_i^2(t) = \{F^{-1}[Y_i(f)]\}^2. \tag{7.12}$$

The mean square value of the response at some time t_1 can be determined by taking an ensemble average of the responses at that time (the asterisks denote complex conjugates);

$$\begin{aligned}
\Psi_y^2(t_1) &= \mathbf{E}\,[y_i^2(t_1)] \\
&= \mathbf{E}[\{F^{-1}[Y_i(f)]\}^2] \\
&= \mathbf{E}\left[\left\{\int_\infty^\infty H(f)X_i(f)e^{j2\pi f t_1}df\right\}^2\right] \\
&= \mathbf{E}\left[\int_\infty^\infty H^*(f_1)X_i^*(f_1)e^{-j2\pi f_1 t_1}df_1 \int_\infty^\infty H(f_2)X_i(f_2)e^{j2\pi t_2 t_1}df_2\right], \tag{7.13}
\end{aligned}$$

$$\Psi_y^2(t_1) = \mathbf{E}\left[\int_\infty^\infty \int_\infty^\infty H^*(f_1)H(f_2)X_i^*(f_1)X_i(f_2)e^{j2\pi t_1(f_2-f_1)}df_1 df_2\right]. \tag{7.14}$$

Since integration is a linear operation and the input transforms are the only random variables, the expectation operator can be brought inside the integral as shown below:

$$\Psi_y^2(t_1) = \int_{-\infty}^{\infty} \int_{-\infty}^{\infty} H^*(f_1)\, H(f_2)\, \mathbf{E}\left[X^*(f_1)\, X(f_2)\right] e^{j2\pi t_1(f_2 - f_1)}\, df_1\, df_2. \qquad (7.15)$$

The quantity $\mathbf{E}\left[X^*(f_1)\, X(f_2)\right]$ is known as the generalized spectral density function ([25], p. 352). This generalized spectral density function is the double Fourier transform of the nonstationary autocorrelation function,

$$\mathbf{E}\left[X^*(f_1)\, X(f_2)\right] = S_x(f_1, f_2) = \int_{-\infty}^{\infty} \int_{-\infty}^{\infty} R_x(t_1, t_2) e^{j2\pi(f_1 t_1 - f_2 t_2)} dt_1 dt_2. \qquad (7.16)$$

The integration in Eq. (7.15) can be slightly simplified by making a change of variables. Let the nonstationary autocorrelation function be defined as follows:

$$R_x(t_1, t_2) = R_x(\tau, t), \qquad (7.17)$$

where

$$\tau = t_2 - t_1 \text{ and } t = \frac{t_1 + t_2}{2}.$$

The generalized spectral density function becomes

$$S_x(f, g) = \int_{-\infty}^{\infty} \int_{-\infty}^{\infty} R_x(\tau, t) e^{-j2\pi(f\tau + gt)} d\tau dt, \qquad (7.18)$$

and the time varying response becomes (because $\tau = 0$ for a mean square value)

$$\Psi_y^2(t) = \int_{-\infty}^{\infty} \int_{-\infty}^{\infty} H^*(f) H(g) S_x(f, g) e^{-j2\pi gt} df dg. \qquad (7.19)$$

Instantaneous Power Spectral Density Function

Another nonstationary spectral approach that has been used is the instantaneous power spectra [43]. The instantaneous power spectrum is a single Fourier transform of the nonstationary autocorrelation function;

$$S_x(f, t) = \int_{-\infty}^{\infty} R_x(\tau, t) e^{-j2\pi f\tau} d\tau. \qquad (7.20)$$

The time-varying, mean square value is the integral of the instantaneous power spectra over all frequencies,

$$\Psi_y^2(t) = \int_{-\infty}^{\infty} S_x(f, t) df. \qquad (7.21)$$

The total energy up to time t_1 is found as follows:

$$E(t_1) = \int_{-\infty}^{t_1} \int_{-\infty}^{\infty} S_x(f, t) \, df \, dt. \tag{7.22}$$

The time-varying and time-averaged power spectra sometimes used for nonstationary vibration analysis [25, 44], are not very satisfactory for computing the time-varying mean square responses of systems to shock inputs because the results of these techniques become highly dependent on the analog filters employed when the signals are transient.

Peak Response Values

A description of the peak response of a system to "random transients" must be stated in probabilistic terms. Generally, the probability of exceeding some amplitude in some fixed period of time is an item of prime interest. Analytical techniques have not yet been developed to provide a general solution to this problem from input data. However, there have been a few cases studied analytically [45, 46]. There have also been a few cases studied empirically [47–49].

In all of these studies, the system considered is a simple, second order mechanical oscillator, and the input is assumed to be white noise multiplied by some simple time function. Typical results from Ref. 48 are reproduced in Figs. 7.9, 7.10, and 7.11 for deterministic time functions of a step function, a boxcar function of duration $\Delta\tau$, and half-sine function of duration $2\Delta\tau$. The quantity β is the ratio of the maximum instantaneous value to the rms value of the stationary noise. This peak-to-rms value is plotted against dimensionless time quantities. The undamped natural frequency of the simple oscillator is f_n. The ratio of the undamped natural frequency to the half-power bandwidth of the oscillator is the value Q. Both T and $\Delta\tau$ are time values. The former is the elapsed time for observing the system response to white noise. The latter is the effective time duration of the pulsed excitation.

7.3 Other Decompositions

Although decomposition of a time history has traditionally been performed in terms of trigonometric functions, there is no reason why expansions cannot be expressed by other mathematical functions. In particular, two approaches which have been used are
- Expansion in terms of orthogonal polynomials
- Expansion in terms of exponentials.

Each of these methods will be discussed in turn.

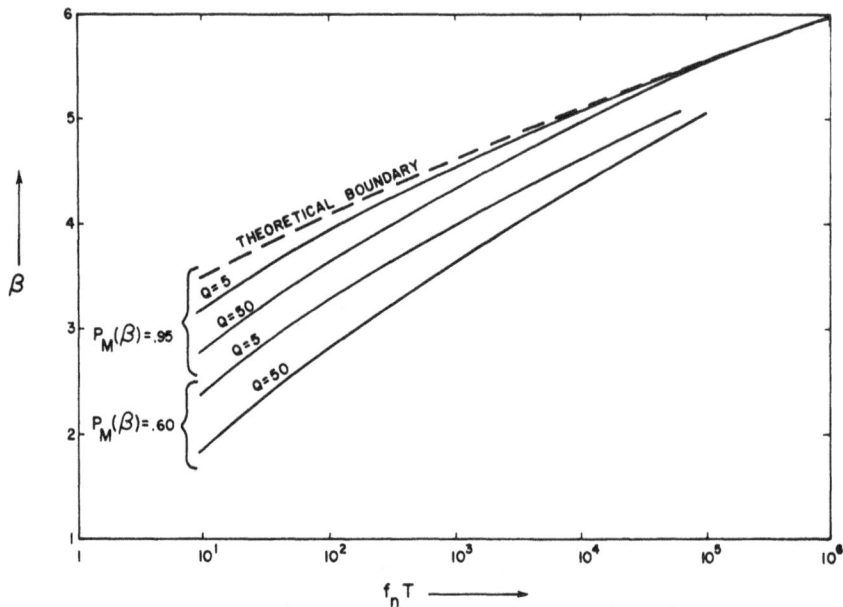

Fig. 7.9. Experimental results for the peak-to-rms response of the mechanical oscillator excited by stationary white noise (β vs f_nT).

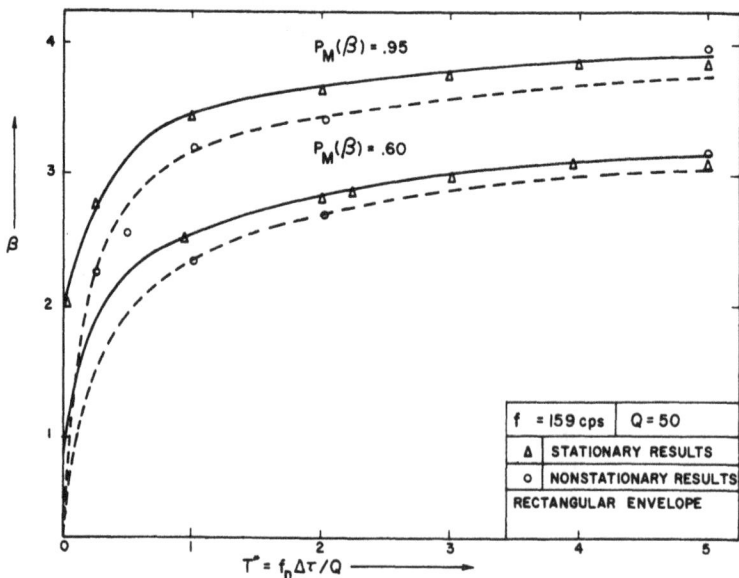

Fig. 7.10. Ratio of peak-to-rms response of the mechanical oscillator β vs the dimensionless time parameter T^* for stationary and pulsed random excitation, rectangular envelope, $Q = 50$.

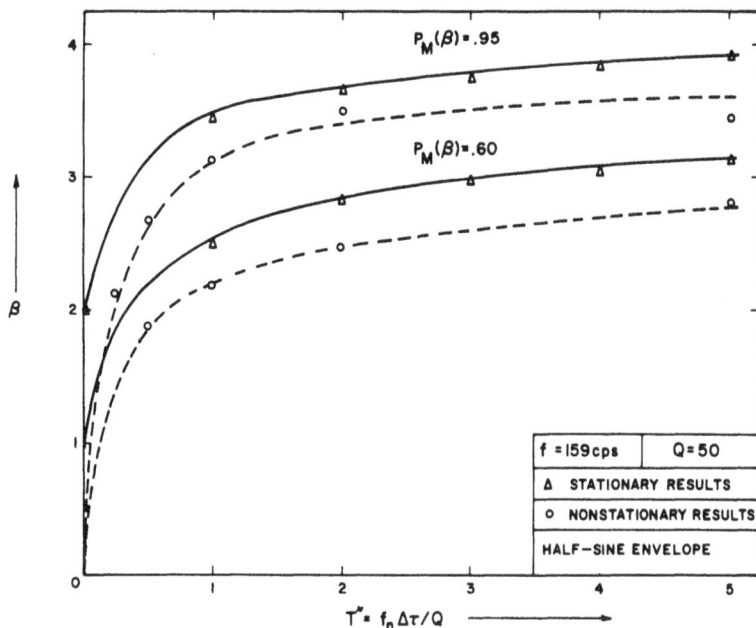

Fig. 7.11. Ratio of the peak-to-rms response of the mechanical oscillator vs the dimensionless time parameter T* for stationary and pulsed random excitation, half-sine envelope, Q = 50.

Orthogonal Polynomials

Orthogonal functions are closed sets of functions for which the following relationship holds:

$$\int_a^b \phi_m(x)\phi_n(x)dx = \begin{cases} 0, & m \neq n \\ A_m, & m = n, \end{cases} \tag{7.23}$$

where the $\phi(x)$ are orthogonal over the interval (a, b). In particular, if all $A_m = 1$, the functions are said to be *orthonormal*.

Examples of orthogonal functions are the sines and cosines used in Fourier analysis. This can be seen from the following relationships:

$$\int_{-\pi}^{\pi} \sin kx \sin \ell x \, dx = \begin{cases} 0, & k \neq \ell \\ \pi, & k = \ell \end{cases} \tag{7.24}$$

$$\int_{-\pi}^{\pi} \cos kx \cos \ell x \, dx = \begin{cases} 0, & k \neq \ell \\ \pi, & k = \ell \neq 0 \\ 2\pi, & k = \ell = 0. \end{cases} \tag{7.25}$$

The decomposition of a time history in terms of orthogonal polynomials is stated by

$$x(t) \approx y(t) = \sum_{i=0}^{M} C_i \phi_i(t), \tag{7.26}$$

where $y(t)$ is an approximation to $x(t)$,

$$C_i = \int_a^b w(t)\phi_i(t)x(t)\,dt, \tag{7.27}$$

and $w(t)$ is some weighting function.

The advantages in using orthogonal polynomial expansions are twofold. First of all, for the interval specified and a given degree, this type of expansion approximates the measured time history with the minimum mean square error. In other words, for some degree M, the expression

$$\int_a^b w(t)[x(t) - y(t)]^2 dt$$

is minimized when

$$y(t) = \sum_{i=0}^{M} C_i \phi_i(t). \tag{7.28}$$

The other advantage in using orthogonal expansions is that each of the functions $\phi_i(t)$ is independent of all others. Because of this independence, the response of a linear system may be determined individually for each component and then summed to define the total response. Also, to add more components to the expansion, one need only compute the required additional terms without recalculating all lower order terms.

While there are many different types of orthogonal polynomials which can be used, two appear particularly appropriate for transient analysis because of their exponential weighting functions and infinite orthogonality intervals.

The first type is the *LaGuerre* polynomial. In this case

$$w(t) = e^{-\alpha t} \tag{7.29}$$

and the polynomial is defined by

$$L_i(t) = e^{\alpha t} \frac{d^i}{dt^i} (t^i e^{-\alpha t}). \tag{7.30}$$

The C_i used in the approximation may be determined from the relationship

$$C_i = \frac{\alpha}{(i!)^2} \int_0^\infty e^{-\alpha x} x(t) L_i(t)\,dt. \tag{7.31}$$

Then

$$x(t) \approx \sum_{i=0}^{M} C_i L_i(t). \tag{7.32}$$

LaGuerre expansions have been investigated for certain types of transients by Quazi [50]. He concluded that, at least for two analytical time histories, the LaGuerre method required fewer terms than Fourier expansions for the same accuracy.

The second type of polynomial is orthogonal over the doubly infinite interval $(-\infty, \infty)$. In this case, the polynomial is known as a Hermitian polynomial. The weighting function is

$$w(t) = e^{-\alpha^2 t^2}, \tag{7.33}$$

and the polynomial is defined by

$$H_i(t) = -(\alpha)^{-i} e^{\alpha^2 t^2} \frac{d^i}{dt^i} (e^{-\alpha^2 t^2}). \tag{7.34}$$

The coefficients of the expansion are obtained from

$$C_i = \frac{\alpha}{2^i i! \sqrt{\pi}} \int_{-\infty}^{\infty} e^{-\alpha^2 t^2} H_i(t) x(t) dt, \tag{7.35}$$

and the expansion of $x(t)$ in terms of $H_i(t)$ is

$$x(t) \approx \sum_{i=0}^{M} C_i H_i(t). \tag{7.36}$$

Further investigation is needed to determine the utility of these methods as applied to shock data.

Exponential Expansion

When analyzing data which decay with time, a natural approach is to assume that the excitation can be approximated by an exponential series of the form

$$x(t) \approx a_1 e^{b_1 t} + a_2 e^{b_2 t} + \ldots + a_n e^{b_n t} \tag{7.37}$$

or, equivalently

$$x(t) \approx a_1 \mu_1' + a_2 \mu_2' + \ldots + a_n \mu_n', \tag{7.38}$$

where

$$\mu_i = e^{b_i}.$$

Note that if $b_i = j\beta_i$, then the expansion becomes

$$x(t) \approx a_1 e^{j\beta_1 t} + a_2 e^{j\beta_2 t} + \ldots + a_n e^{j\beta_n t}, \tag{7.39}$$

where

$$e^{j\beta t} = \cos \beta t + j \sin \beta t. \tag{7.40}$$

This is the equivalent of fitting a Fourier series to the data at the frequencies $\beta_1, \beta_2, \ldots \beta_n$. The important item to realize is that, in general, both sets of coefficients (the a_i's and the b_i's) are unknown and must be obtained from the excitation time history. A procedure for determining these coefficients is described by Hildebrand [51] and is called Prony's method. The method is digital in nature, so it will be assumed that the excitation has been digitized to produce samples at some constant time increment. Therefore,

$$x_i = x(i\Delta t), \qquad i = 0, 1, 2, \ldots, N.$$

Equation (7.38) indicates that the solution for the a_i's is linear; however, the μ_i's are definitely nonlinear. Since it is difficult to determine non-linear coefficients by normal methods, the equivalent of a transformation of variables must be made. This is performed by defining the μ_i as roots of an algebraic equation

$$\mu^n - \alpha_1\mu^{n-1} - \alpha_2\mu^{n-2} \ldots - \alpha_{n-1}\mu - \alpha_n = 0, \tag{7.41}$$

so that the left-hand side of the equation is identical to $(\mu - \mu_1)$ $(\mu - \mu_2) \ldots (\mu - \mu_n)$. It is now necessary to determine the values of the α_i. This can be accomplished by solving the set of n linear equations

$$\alpha_1 x_{n-1} + \alpha_2 x_{n-2} + \ldots + \alpha_n x_0 = x_n$$

$$\alpha_1 x_n + \alpha_2 x_{n-1} + \ldots + \alpha_n x_1 = x_{n+1}$$

$$\cdot \quad \cdot \quad \cdot \qquad \qquad \cdot \quad \cdot \quad \cdot$$

$$\alpha_1 x_{2n-2} + \alpha_2 x_{2n-3} + \ldots + \alpha_{n-1} x_{n-1} = x_{2n-1}. \tag{7.42}$$

Equation (7.42) as shown assumes that $N = 2n$. If $N > 2n$, the system is overdetermined and must be solved by least squares procedures. In any case, the α's are used in conjunction with Eq. (7.41) in order to determine the μ_i's. This determination will usually require some iterative root-finding, especially if n is greater than four.

Since the defining Eq. (7.38) is linear in the a_i's, it is now only necessary to solve the following set of linear equations simultaneously to complete the procedure:

$$a_1 + a_2 + \ldots + a_n = x_0$$

$$a_1\mu_1 + a_2\mu_2 + \ldots + a_n\mu_n = x_1$$

$$a_1\mu_1^2 + a_2\mu_2^2 + \ldots + a_n\mu_n^2 = x_2$$

$$\cdot \quad \cdot \quad \cdot \qquad \qquad \cdot \quad \cdot \quad \cdot$$

$$a_1\mu_1^{n-1} + a_2\mu^{n-1} + \ldots + a_n\mu^{n-1} = x_{n-1}. \tag{7.43}$$

Again, as with the solution of Eq. (7.42) for the a_i's, this set may be solved uniquely for $N=n$ or by least squares procedures for $N>n$.

If the transient to be analyzed is assumed to be a set of damped sinusoids and cosinusoids at varying frequencies, then Eq. (7.37) may be written as

$$x(t) = A_1 e^{\gamma_1 t} \sin \omega_{d_1} t + A_2 e^{\gamma_2 t} \cos \omega_{d_2} t + \ldots + A_1 e^{\gamma_k t} \sin \omega_{d_k} t, \quad (7.44)$$

where the γ_1 are the damping factors, the A_i are the amplitudes, and the ω_{d_i} are the damped frequencies. Equation (7.37) may be solved by Prony's method to obtain the γ_i, the A_i, and ω_{d_i}. To calculate the damping ratios ζ_i and the undamped natural frequencies ω_{n_i}, it is only necessary to rewrite Eq. (7.37) as

$$x(t) = A_1 e^{-\zeta_1 \omega_{n1} t} \sin (\omega_{n_1} t \sqrt{1-\zeta_1^2}) + A_2 e^{-\zeta_2 \omega_{n2} t} \cos (\omega_{n_2} t \sqrt{1-\zeta_2^2}) + \ldots$$
$$+ A_k e^{-\zeta_k \omega_{nk} t} \sin (\omega_{n_k} t \sqrt{1-\zeta_k^2}), \quad (7.45)$$

where

$$A_1 = \frac{a_i}{\sin (\cos) \omega_{d_i} t}$$

$$\omega_{n_i} = \sqrt{\omega_{d_i}^2 + \gamma_i^2} \quad (7.46)$$

$$\zeta_i = \frac{\gamma_i}{\sqrt{\omega_{d_i}^2 + \gamma_i^2}}. \quad (7.47)$$

While this method does provide an analytic expression for the excitation, it suffers from two basic problems. The first problem occurs because it is possible for several different sets of coefficients to provide equally good results as far as approximating the time history is concerned, especially if the signal is corrupted by noise. Because of the nonlinearity of the approximation, small changes in the data can cause significant changes in the coefficients. The second problem is that it is difficult to determine the number of exponentials which best approximates a given excitation. Since this must be known prior to the analysis, it is necessary that the engineer have a good estimate of the number of significant components contained in the excitation before utilizing this procedure.

7.4 Extensions of the Basic Shock Spectrum Concept

The basic shock spectrum concept requires that the physical system whose response is being computed must be accurately represented by a second order linear oscillator. It further requires that only the maximum response value of this system need be measured. This simple concept has been applied to a number of systems that do not meet the above requirement. It has been used to analyze the response of certain nonlinear systems, certain multiple degree-of-freedom systems (the second order mechanical oscillator is frequently called a single degree-of-freedom system), and collisions between two second order systems. In addition, the peak probability density function of the second order system has been computed to study the cumulative fatigue damage potential of shocks. These variations of the fundamental shock spectrum concept are described in this section.

Special Normalization

Crede [52] proposes that the shock spectrum should be normalized by dividing the maximum acceleration response by the product of the velocity change of the shock and the undamped natural frequency of the second order mechanical oscillator. The advantage claimed is that it makes the spectrum relatively insensitive to the shape of the shock. It does this by dividing the normal shock spectra by frequency, so that the low frequency portion is emphasized and the high frequnecy portion deemphasized. Since most shock spectra differ more at high frequencies than at low, this reduces these differences. Normalizing by the velocity change causes the low frequency asymptote to be unity. This can be seen from Eq. (7.5), where the shock spectrum, for a step change in velocity, is given by

$$a_{max} = \Delta V 2\pi f_n.$$

Therefore,

$$C = \frac{a_{max}(2\pi f_1)}{\Delta V 2\pi f_n} \approx 1,$$

at low frequencies where the response is not as sensitive to waveform details. Figure 7.12 shows the spectra of several classical pulses of duration T_0 renormalized in this manner. Since the resultant spectra are relatively similar, a single curve can be used in conjunction with the velocity change to predict the peak response. Techniques for applying this spectra are discussed in Ref. 53.

Analysis of Multiple Degree-of-Freedom Systems

The use of shock spectrum procedures in the determination of the response of complex structures to a shock is fraught with controversy. For the shock spectrum to have real meaning in terms of stresses or loads generated in a multiple degree-of-freedom system, the response must be limited primarily to one mode. Such a situation can occur only if the exciting force contains significant energy in the vicinity of only one of the system modal frequencies. Unfortunately, many transients are broadband in nature, containing energy spread over a considerable frequency range. This usually causes excitation of several system modes simultaneously. As a result, the true system response will con-

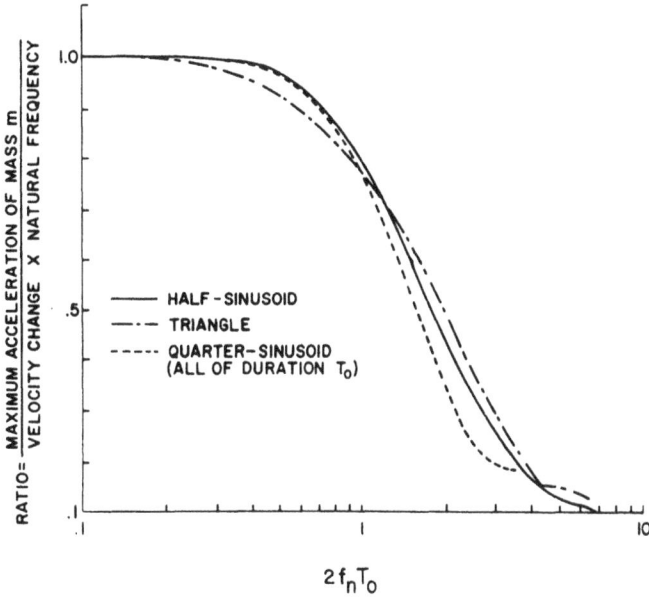

Fig. 7.12.   Shock spectrum as normalized by Crede.

sist of the algebraic sum of the responses of each of the excited modes. The above statement assumes that no loading or feedback exists in the system.

The shock spectrum is inherently a single degree-of-freedom concept. Furthermore, it contains information concerning only the peak response amplitude at each of the specified system natural frequencies. No information as to the time of occurrence of each of the peak responses is maintained. Finally, the entire set of response peaks making up the shock spectrum is computed for the same damping factor, whereas in the usual multiple degree-of-freedom system, each system mode has a different damping factor associated with it. Because of all these reasons, meaningful estimates of the true system response may be obtained if the system responds in one and only one mode.

For any multiple degree-of-freedom system, the response to an excitation may be written as

$$\mathbf{R}(t) = \sum_{n=1}^{N} a_n \boldsymbol{\phi}^{(n)} \int_0^t h_n(t-\tau) f(\tau) \, d\tau, \qquad (7.48)$$

where

$\boldsymbol{\phi}^{(n)}$ is the nth system modal vector
a_n is the modal participation constant for mode n
$h_n(\tau)$ is the unit impulse response function for mode n

$f(\tau)$ is the exciting force
N is the number of system modes.

When the response is primarily in one mode (mode m), Eq. (7.48) reduces to

$$\mathbf{R}(t) = \boldsymbol{\phi}^{(m)} \int_0^t h_m(t-\tau) f(\tau) d\tau. \tag{7.49}$$

If the value of the undamped maximum shock spectrum at the frequency corresponding to the mth system mode is defined as $q_{m,\ max}$, then

$$q_{m,\ max} = \max_{t \geq 0} \left[\int_0^t h_m(t-\tau) f(\tau) d\tau \right] \tag{7.50}$$

and, finally,

$$\max_{t \geq 0} \left[\mathbf{R}(t) \right] \approx \boldsymbol{\phi}^{(m)} q_{m,\ max}. \tag{7.51}$$

In this manner the maximum response may be obtained.

In certain isolated cases where the response parameter of interest depends strongly on only one response mode, it may be possible to ignore the responses of other system modes and to consider the system as responding only in this mode. As an example of this, consider the following system as described by Cronin [54].

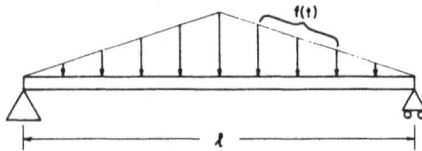

Fig. 7.13. Simply supported beam excited by a symmetrically applied force.

A simply supported beam of length ℓ is excited by a force $f(t)$ as shown in Fig. 7.13. The required response parameter is the deflection of the midpoint of the beam. The symmetry of the excitation implies that only the odd system modes will be excited. Furthermore, the modal participation constants will be proportional to $1/n^2$, where n is the mode number. Therefore, for $n > 1$,

$$a_m \leq 1/9 a_1,$$

implying that all modes other than the first are unimportant so that the required response is basically a function of only the first mode, and from Eq. (7.51),

$$\max_{t \geq 0} \left[\mathbf{R}(t) \right] = c\phi^{(1)} \cdot q_{1,\text{max}}, \tag{7.52}$$

where c is some constant of proportionality.

If the system responds in several modes, it is possible to obtain an upper bound on the true response, provided that knowledge of all excited system modes is available. This bound was derived initially by Biot [1], and by Biot and Bisplinghoff [55] in the following form:

$$\max_{t \geq 0} |R_i| \leq \sum_{n=1}^{N} |a_n| \cdot |\phi_i^{(n)}| \cdot q_{n,\text{max}}, \tag{7.53}$$

where R_i is the response of the ith mass. While this bound gives reasonable results for earthquakes and other transients with relatively long decay times, it is overly conservative for the single pulse type of shock often encountered. The accuracy of the bound also deteriorates as the number of participating modes increases. For this reason, it is best to restrict the number of modes used in computing the bound to only those providing significant responses.

A less conservative bound has been obtained by Fung and Barton [56] for the specific case where the rise time of the shock pulse is large when compared with the half-period of the lowest system mode. In this case, all the excited system modes will follow the excitation very closely. As a result, the peak response of *all* modes will occur at about the same time as the peak of the exciting pulse. Furthermore, they will all respond in the same direction. The bound is then simply

$$\max_{t \leq 0} R_i \leq \sum_{n=1}^{N} a_n \cdot \phi_i^{(n)} \cdot q_{n,\text{max}}. \tag{7.54}$$

It should be evident from the above discussion that only gross information concerning the system responses of complex structures may be obtained by shock spectrum procedures. For a more accurate analysis, it is necessary to determine the phase relationships between the various modal responses. If all the responding modes are known, Eq. (7.48) may be utilized to obtain the required response time history. However, in many cases it is easier to perform the analysis in the frequency domain. In this manner, the convolution is reduced to a multiplication by the various modal frequency response functions. The response time history is then obtained by taking the inverse Fourier transform of the resultant weighted transform of the excitation.

In summary, shock spectrum techniques may be used only to obtain conservative bounds for the system response of a complex structure which is responding in more than one mode. For this type of system, Fourier spectrum techniques will generally provide much more accurate results.

Three-Dimensional Shock Spectrum

The three-dimensional shock spectrum is a method for deriving fatigue damage information from a shock which does not cause catastrophic failure. For example, consider the effects of carrier landing shocks upon the landing gear of an aircraft. In general, it is the cumulative fatiguing effects of many such shocks which eventually cause the failure of the landing gear rather than a single peak situation such as those normally analyzed by direct shock spectrum methods.

Fatigue damage analysis is usually performed by means of S-N curves and the Miner linear rule. An S-N curve shows the typical relation between the stress level S and the number of cycles of stress reversal (at the stress level S) required to cause failure of the system. The Miner linear rule utilizes this curve in determining fatigue failure when stress reversals at differing maximum stress levels occur. For example, a system is subjected to n_1 cycles at a maximum stress level of S_1, n_2 cycles at a level of S_2, etc. As shown in Fig. 7.14, then, the system may be expected to fail if

$$\frac{n_1}{N_1} + \frac{n_2}{N_2} + \frac{n_3}{N_3} + \ldots + \frac{n_m}{N_m} > 1, \tag{7.55}$$

where the N_i are the numbers of cycles required for failure at the stress levels S_i. To perform cumulative-fatigue failure analysis, one must determine the number of stress reversals as a function of the stress levels attained. The three-dimensional shock spectrum provides this information.

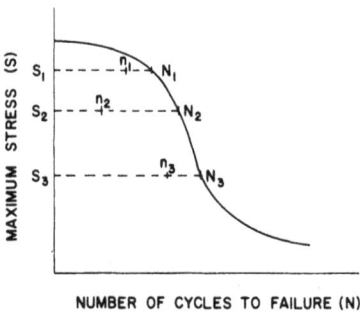

Fig. 7.14. Typical S-N curve.

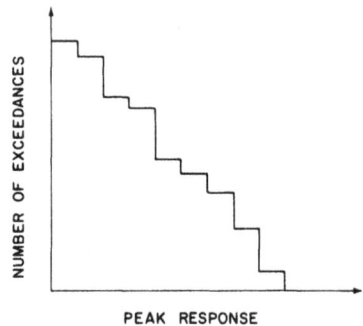

Fig. 7.15. Peak response histogram.

Consider the response time histories obtained when computing the shock spectrum. If all the relative maxima are detected instead of just the maximum value, then a histogram or bar chart showing the number of maxima exceeding various response levels may be drawn. An example of such a histogram is shown in Fig. 7.15. This histogram provides the information needed in determining fatigue damage by Eq. (7.55) for one natural frequency only. If the histograms corresponding to all natural frequencies of interest are computed and a three-dimensional plot of peak response versus the number of exceedances and the natural frequency is drawn, then it would appear as in Fig. 7.16. The result is a surface usually denoted as the response surface. Its intersection with the natural frequency plane of the peak response corresponds to the normal shock spectrum of the system.

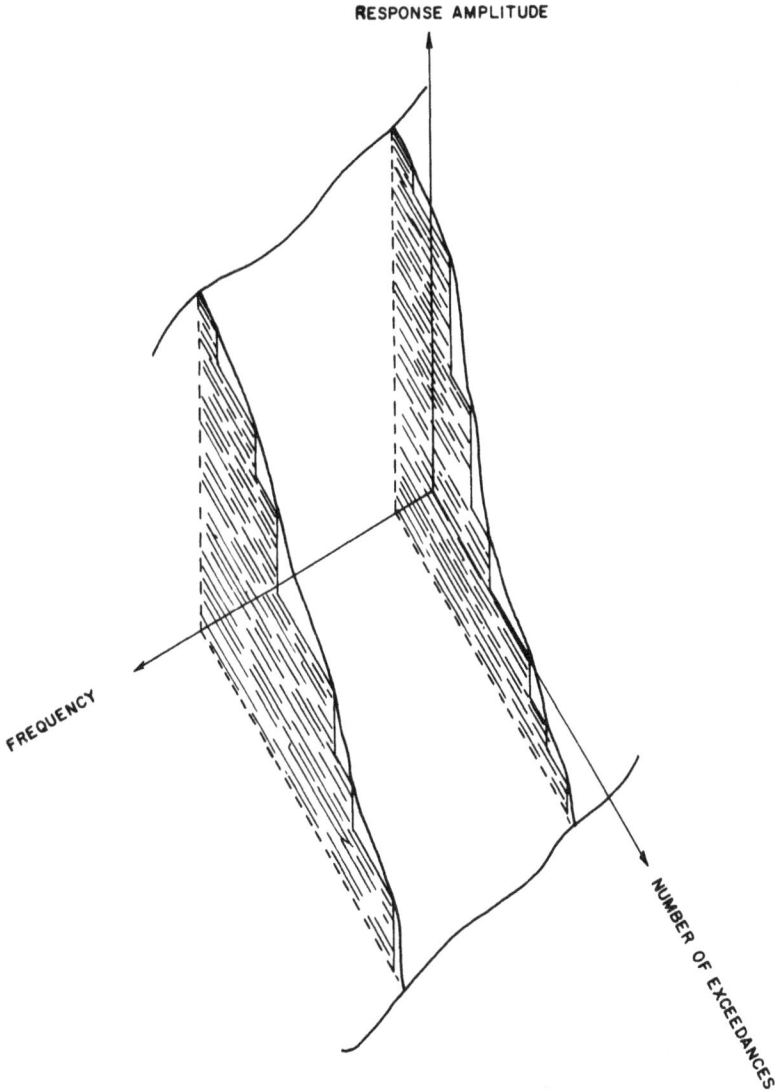

RESPONSE AMPLITUDE

FREQUENCY

NUMBER OF EXCEEDANCES

Fig. 7.16.   Three-dimensional shock spectrum.

The response surface is of more qualitative than quantitative interest, since it is difficult to determine visually the number of stress reversals at various stress levels. Because of this, the individual histograms are used instead of the three-dimensional surface.

This procedure suffers from the same problem as all the response spectrum techniques in that no information is available as to the true system response but only the response of the assumed simple second order system model for the system. Nevertheless, the three-dimensional shock spectrum is a useful concept because it makes it possible to define a laboratory shock test whose fatigue characteristics are similar to the environmental shock. This procedure is detailed in Ref. 57.

The Proximity Spectrum

The proximity spectrum is a recent development due to Schell [58] for extending the shock spectrum concept to certain types of damage or failure which cannot be assessed by the usual shock spectrum procedures. This method provides an evaluation of the damage potential of a shock on equipment where the primary cause of failure is the interaction of internal components rather than the fatiguing of the system.

Two basic types of equipment are prone to this type of failure. In mechanical systems, the failure is due to increased contact between components. This is evidenced usually by friction or collision between various parts of the system. In electromagnetic systems, a change in the proximity of components may cause changes in dielectric strength, loss of insulation resistance, and variations in magnetic or electrostatic field strengths.

If an item of equipment can be approximated by the model shown in Fig. 7.17, then the proximity spectrum may provide valuable insight in its capability to withstand shock. Note that the model consists of a pair of single degree-of-freedom systems mounted on a common base. Each system is independent of the other, with different masses, spring constants, and viscous damping coefficients. As a result, when a shock motion is applied to the left side of the base, each system will respond differently, causing the distance between the two masses to vary as a function of time. The distance D is composed of a static component D_{st} and a dynamic component $\Delta(t)$, which is the difference between the dynamic displacements of the two masses. In equation terms,

$$D(t) = D_{st} + \Delta(t), \tag{7.56}$$

where

$$\Delta(t) = x_2(t) - x_1(t).$$

The $\Delta(t)$ function is also called the proximity criterion. Although the entire time history of Δ is of value in determining the damage potential

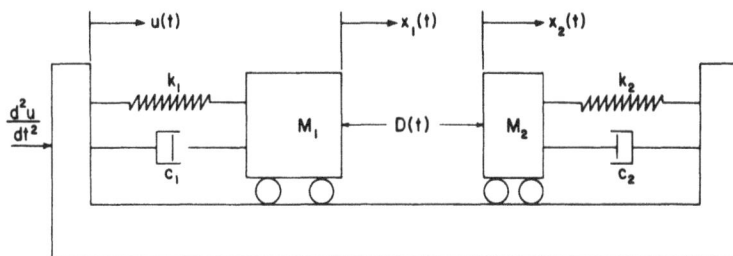

Fig. 7.17. System model for proximity spectrum.

of a shock, the extreme values of Δ contain the most important information. By calculating these extrema for varying ratios of the two natural frequencies f_1 and f_2, a spectrum similar to the shock spectrum is obtained. This is the proximity spectrum.

To compute the proximity time history $\Delta(t)$, the equations of motion of the two masses must be solved simultaneously. These equations are

$$m_1\ddot{x}_1 + c_1(\dot{x}_1 - \dot{u}) + k_1(x_1 - u) = 0$$

and

$$m_2\ddot{x}_2 + c_2(\dot{x}_2 - \dot{u}) + k_1(x_2 - u) = 0. \tag{7.57}$$

Define the relative displacements of the two masses m_1 and m_2 by

$$\delta_1 = x_1 - u \qquad \text{and} \qquad \delta_2 = x_2 - u. \tag{7.58}$$

Then Eq. (7.57) becomes

$$m_1\ddot{\delta}_1 + c_1\dot{\delta}_1 + k_1\delta = -m_1\ddot{u}(t)$$

and

$$m_2\ddot{\delta}_2 + c_2\dot{\delta}_2 + k_2\delta = -m_2\ddot{u}(t). \tag{7.59}$$

Replacing the m, k, and c terms by their equivalents in terms of the natural circular frequency (ω_n) and the damping factor produces

$$\ddot{\delta}_1 + 2\zeta\omega_{n_1}\dot{\delta}_1 + \omega_{n_1}\delta_1 = -\ddot{u}(t)$$

$$\ddot{\delta}_2 + 2\zeta\omega_{n_2}\dot{\delta}_2 + \omega_{n_2}\delta_2 = -\ddot{u}(t), \tag{7.60}$$

or

$$\delta_1 = -\frac{\ddot{u}(t) + \ddot{\delta}_1}{\omega_{n_1}^2} - \frac{2\zeta\dot{\delta}_1}{\omega_{n_1}}$$

$$\delta_2 = -\frac{\ddot{u}(t) + \ddot{\delta}_2}{\omega_{n_2}^2} - \frac{2\zeta\dot{\delta}_1}{\omega_{n_2}}, \tag{7.61}$$

since

$$\Delta = x_2 - x_1 = (x_2 - u) - (x_1 - u) = \delta_2 - \delta_1. \tag{7.62}$$

Then

$$\Delta(t) = \frac{\ddot{u}(t) + \ddot{\delta}_1}{\omega_{n_1}^2} - \frac{\ddot{u}(t) + \ddot{\delta}_2}{\omega_{n_2}^2} + 2\zeta \left(\frac{\dot{\delta}_1}{\omega_{n_1}} - \frac{\dot{\delta}_2}{\omega_{n_2}} \right). \tag{7.63}$$

Note that, throughout the derivation of Eq. (7.63), it has been assumed that both systems have identical damping factors. Schell [59] feels that it is sufficient to perform the analysis in this manner, utilizing values of 0.005 and 0.1 for ζ.

The results are usually presented in a normalized manner with the ordinate of the plot being Δmax or $-\Delta$min divided by $E_p\tau^2$, where τ is the duration of the shock pulse and E_p is the peak amplitude of the excitation. The normalized abscissa is simply τf_2, and a family of curves for different values of τf_1 is plotted on each graph. Figure 7.18 shows the proximity spectrum for a terminal-peak sawtooth pulse. For this particular case, the negative and positive spectra are identical, but in general this relationship will not hold true.

Figure 7.19 [58] indicates the proximity spectrum of a square wave pulse. For values of τf_1 below 0.9, the spectra are quite simple. Above 0.9, they exhibit peaks and valleys and sudden trend reversals. Schell states that these reversals are due to the fact that the extrema occurring during the initial pulse and those occurring during the residual period display opposite trends. Therefore, a crossover point exists at which the extrema obtained from the initial period become greater or less than those obtained from the residual period. Since only the largest and smallest of these extrema are utilized in the proximity spectrum regardless of their time of occurrence, a sudden trend reversal occurs in the curve at this changeover point.

Schell [58] gives proximity spectra for a considerable number of theoretical pulse shapes, but the concept is too new to have been utilized on measured shock pulses and real systems. As a result, it is not possible to estimate its eventual use at this time.

Nonlinear Shock Spectra

One of the fundamental assumptions in shock spectral analysis is that the physical system can be represented by a *linear* second order system. There are no truly linear systems. The practical problem usually becomes one of assuring that the deviations from linearity are sufficiently slight so that the shock spectrum technique can be validly used. However, there are cases where the system is clearly nonlinear, and deviations from linearity are major rather then minor. Several of the basic concepts of the shock spectrum have been applied to analyze several types of nonlinearities. The basic concepts retained are that the maximum response of a second order system is indicative of the damage

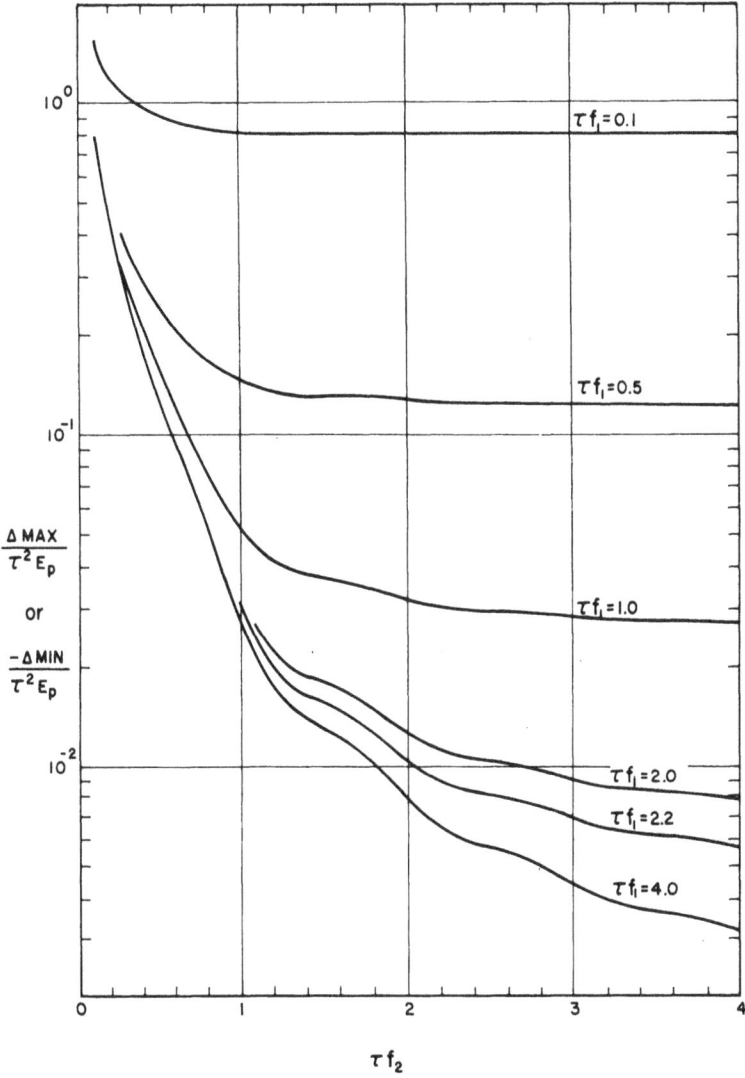

Fig. 7.18. Undamped proximity spectrum of a terminal-peak sawtooth excitation function.

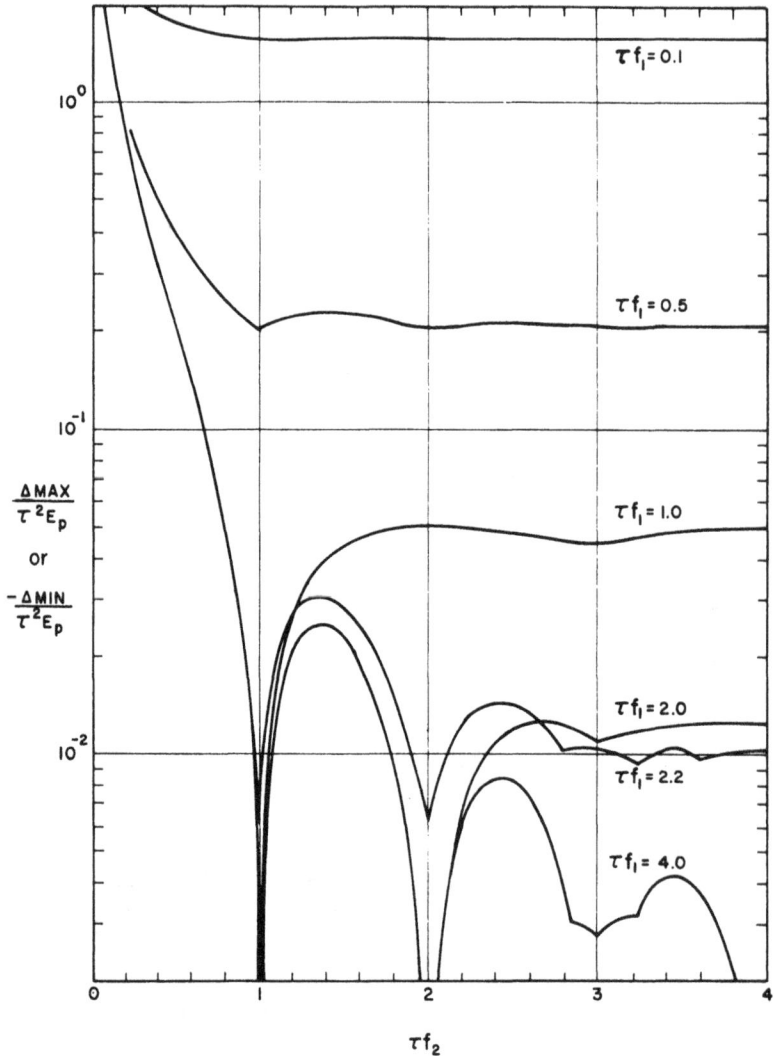

Fig. 7.19. Undamped proximity spectrum of a square-wave excitation function.

potential of the shock, and that dynamic responses can be converted to static values.

The nonlinear shock spectrum is obtained by replacing one of the linear elements with a nonlinear one, applying the excitation, and computing the maximum response. Examples of the shock spectra response to certain pulses of second order systems having hardening and softening spring nonlinearities are contained in Refs. 60 and 61. Ref. 62 considers the strain hardening dead zone and bilinear stiffness nonlinearities.

SYMBOLS

A	Constant, usually denoting amplitude or area
$A_{eq}(t)$	Equivalent static acceleration time history
a	Constant
a_n	Fourier series cosine coefficient, or modal participation constant
B	Bandwidth in hertzes or spatial flux density
b	Constant
b_n	Fourier series sine coefficient
C	Capacitance
C_k	Real part of the discrete Fourier transform
c	Viscous damping coefficient
c_n	Fourier series modulus
D	Distance
D_n	Normalized Fourier series
D_{st}	Static distance
E	Energy
$E[\ ]$	Mathematical expectation
$E(f)$	Fourier transform of $e(t)$
$e(\%)$	Expected error (percentage)
$e(t)$	Voltage time history
$F[\ ]$	Fourier transform
$F^{-1}[\ ]$	Inverse Fourier transform
$F(t)$	Exciting or input force time history
$F_c(t)$	Damping force time history
$F_k(t)$	Spring force time history
$F_m(t)$	Inertial force time history
$f(x)$	Function of x

187

f	Frequency in hertzes
$f(\tau)$	Exciting force time history
f_n	Undamped natural frequency in hertzes
f_r	Response frequency in hertzes
f_s	Sampling rate in samples/sec
$G_{xx}(f)$	Spectral density function
$G_{xy}(f)$	Cross spectral density function
g	Gravitational constant
$g(\ )$	Function of the term in the parentheses
H	Magnetic force
$H(f)$	Frequency response function
$|H(f)|$	Gain factor (Fourier transform of $h(t)$)
$H(s)$	Transfer function (Laplace transform of $h(t)$)
$H(z)$	Z-transform of $h(t)$
H_i	Gain factor at the ith frequency
$h(t)$	Unit impulse response or weighting function
h_i	Discrete unit impulse response function
$I(f)$	Fourier transform of $i(t)$
$Im[\ ]$	Imaginary part of a complex quantity
i	Constant
$i(t)$	Current time history
j	Imaginary number ($\sqrt{-1}$)
$K(\ )$	Kernel
$K.E.$	Kinetic energy
k	Constant, usually denoting spring constant
L	Inductance
$\mathscr{L}[\ ]$	Laplace transform
$\mathscr{L}^{-1}[\ ]$	Inverse Laplace transform
ℓ	Length
M	Constant, usually denoting the ratio of sampling frequency to response frequency
m	Mass

N	Constant denoting an integer number, the number of turns in an inductor, or the number of samples in a record
n	Constant
p	Linear operator
p_0, p_1	Nonrecursive filter weights
$P.E.$	Potential energy
$Q(f)$	Fourier transform of $q(t)$
Q_k	Imaginary part of the discrete Fourier transform
$q(t)$	Charge time history
q_1, q_2	Recursive filter weights
R	Resistance
$\mathbf{R}(t)$	Vector response time history
$Re[\ ]$	Real part of complex quantity
$R_x(\tau, u)$	Nonstationary autocorrelation function
$R_{xy}(\tau)$	Crosscorrelation function
r	Radius
$r_1, r_2,$ etc.	Roots of an equation
$S_x(f_1, f_2)$	Generalized spectral density function
$S_x(f, t)$	Instantaneous power spectral density function
s	Laplace transform variable or sample standard deviation
$s(f)$	Spectral standard deviation
$s(t)$	Temporal standard deviation
$T[\ ]$	Linear integral transform
T	Record length
t	Time; independent variable of time history
$u(t)$	Unit step function
$V(t)$	Pseudo velocity time history
$V(s)$	Laplace transform of relative velocity
W	Complex exponential $\left(e\dfrac{-j2\pi}{N}\right)$
$|\bar{X}|$	Mean absolute value
$|X(f)|$	Fourier transform of $x(t)$

$X(f)$	Modulus of a Fourier transform
$\hat{X}(f)$	Estimate of a Fourier transform
$X^m(f)$	Measured value of a Fourier transform
$\overline{X}(f)$	Average Fourier transform
$X_c(f)$	Fourier cosine transform
$X_s(f)$	Fourier sine transform
$X_{c1}(f)$	One-sided Fourier cosine transform
$X_{s1}(f)$	One-sided Fourier sine transform
X_k	Discrete Fourier transform
$\mathscr{X}(f)$	Fourier series
$X(s)$	Laplace transform of $x(t)$
$x(t)$	Input, or excitation, time history
$\bar{x}(t)$	Average time history
x_i	Discrete time series
$Y(f)$	Fourier transform of $y(t)$
$Y(s)$	Laplace transform of $y(t)$
$y(t)$	Output, or response, time history
y_{11}	Input admittance
y_{12}	Reciprocal of the transfer impedance
y_{22}	Output admittance
Z	Impedance
$Z[\ ]$	Z-transform
$Z(s)$	Laplace transform of $\xi(t)$
Z_{11}	Input impedance
Z_{12}	Reciprocal of the transfer admittance
Z_{22}	Output impedance
z	z-transform variable
α	Constant
β	Constant
$\Delta(t)$	Proximity criterion
Δf	Frequency interval
Δt	Time interval

ΔV	Velocity change
Δx	Quantizing increment
$\delta(t)$	Delta function
ϵ_p	Peak response error
ϵ_f	Frequency shift error
ζ	Critical damping ratio
θ	Phase angle
$\theta(f)$	Phase angle of a Fourier transform, or the system phase factor
θ_i	Phase factor at the ith frequency
Λ	Pole of frequency response function
λ	Scan rate in Hz/sec, or frequency
μ	Permeability
μ_i	Roots of an algebraic equation
$\xi(t)$	Relative displacement time history
$\dot{\xi}(t)$	Relative velocity time history
$\ddot{\xi}(t)$	Relative acceleration time history
σ	Real part of Laplace-transform variable
τ	Time delay
$\tau[\ ]$	Logarithmic transform
ϕ	Phase shift
$\phi(x)$	Orthogonal function
ϕ_n	Fourier series phase shift
ϕ_i	System phase shift at the ith frequency
$\boldsymbol{\phi}(n)$	System nth modal vector
Φ	Flux
Ψ^2	Mean square value
$\sqrt{\Psi^2}$	Root mean square value
ω	Frequency in rad/sec
ω_d	Damped natural frequency in rad/sec
ω_n	Undamped natural frequency in rad/sec
$*$	Complex conjugate

REFERENCES

1. M. A. Biot, "A Mechanical Analyzer for the Prediction of Earthquake Stresses," *Bull. Seismol. Soc. Amer.*, **31**, 151–171 (1941).
2. R. H. Rector, "Some Shock Spectra Comparisons Between the ATMX 600 Series Railroad Cars and a Railroad Switching Shock Test Facility," *Shock and Vibration Bull.* **30**, Part 3, 138–164, Shock and Vibration Information Center (Feb. 1962).
3. J. H. Green, C. R. Nisewanger, and R. T. Koyamatsu, "The Response of Missile Components to Water Entry Shock," *Shock and Vibration Bull.* **26**, Part 2, 21–27, Shock and Vibration Information Center (Dec. 1958).
4. V. S. Noonan and W. E. Noonan, "Structural Response to Impulsive Loading (Pyrotechnic Devices)," *Shock and Vibration Bull.* **35**, Part 6, 265–284, Shock and Vibration Information Center (Apr. 1966).
5. R. E. Reisler, "The Mechanical Self-Recording Pressure Time Gage – A Useful Device for the Acquisition of Air Blast Data From Nuclear and Large HE Detonation," *Shock and Vibration Bull.* **28**, Part 3, 99–112, Shock and Vibration Information Center (Sept. 1960).
6. O. Heaviside, *Electrical Papers*, Macmillan, New York, 1892.
7. H. B. Dwight, *Tables of Integrals and Other Mathematical Functions*, Macmillan, New York, 1961.
8. M. F. Gardner and J. L. Barnes, *Transients in Linear Systems*, John Wiley and Sons, New York, 1942.
9. G. A. Campbell and R. M. Foster, *Fourier Integrals*, Van Nostrand Co., Princeton, N.J., 1948.
10. *Tables of Integral Transform*, Vol. I, Bateman Manuscript Project, McGraw-Hill Book Co., New York, 1954.
11. H. S. Carslaw, *Introduction to the Theory of Fourier Series and Integrals*, Dover Publications, New York, 3d edition, rev. 1930, 223.
12. L. E. Dickson, *First Course in the Theory of Equations*, John Wiley and Sons, New York, 1922.
13. R. V. Churchill, *Introduction to Complex Variables and Applications*, McGraw-Hill Book Co., New York, 1948.
14. A. Papoulis, *The Fourier Integral and Its Application*, McGraw-Hill Book Co., New York, 1962.
15. W. R. Lepage and S. Seely, *General Network Analysis*, McGraw-Hill Book Co., New York, 1952.

16. M. R. C. Trubert, "An Analog Technique for the Equilization of Multiple Electromagnetic Shakers for Vibration Testing," *Spacecraft and Rockets*, 5, No. 12 (Dec. 1968).

17. M. A. Biot, "Theory of Elastic Systems Vibrating Under Transient Impulse with an Application to Earthquake-Proof Building," *Proc. Natl. Acad. Sci.* 19:262–268 (1933).

18. S. Rubin, "Concepts In Shock Data Analysis," Ch. 23, *Shock and Vibration Handbook*, Vol. 2 (C. M. Harris and C. E. Crede, Eds.), McGraw-Hill Book Co., New York, 1961.

19. M. Gertel and R. Holland, "Analysis of Shock Records Using a Digital Computer," Frankford Arsenal Report R-1763, DDC AD465410, May 1965.

20. G. W. Painter and J. J. Parry, "Simulating Flight Environment Shock on an Electrodynamic Shaker," *Shock and Vibration Bull.* 33, Part 3, 85–96, Shock and Vibration Information Center, 1964.

21. G. J. O'Hara, "Impedance and Shock Spectra," *J. Acoust. Soc. Am.* 31, 1300–1303 (Oct. 1959).

22. J. P. Walsh and R. E. Blake, "The Equivalent Static Acceleration of Shock Motions," *Proc. SESA*, 6, No. 2, pp. 150–158 (1949).

23. R. F. Carls, "Shock Spectrum Analyzer for General Laboratory Use," *Proc. 1966 Ann. IES.*

24. C. L. Johnson, *Analog Computer Techniques*, McGraw-Hill Book Co., New York, 1956.

25. J. S. Bendat and A. G. Piersol, *The Measurement and Analysis of Random Data*, John Wiley and Sons, New York, 1966.

26. A. A. Kharkevich, *Spectra and Analysis*, Consultants Bureau, New York, 1960.

27. C. E. Shannon, "Communication in The Presence of Noise," *Proc. IRE*, 37, No. 1, 10 (Jan. 1949).

28. G. Goertzel, "An Algorithm for the Evaluation of Finite Trigometric Series," *Am. Math. Monthly*, 65, No. 1, 34–35 (Jan. 1958).

29. J. W. Cooley and J. W. Tukey, "An Algorithm for the Machine Calculation of Complex Fourier Series," *Math. Computation*, 19, 297 (Apr. 1965).

30. W. M. Gentleman and G. Sande, "Fast Fourier Transforms for Fun and Profit," *Proc. Fall Joint Comput. Conf.*, 1966, 563–578.

31. C. Bingham, M. D. Godfrey, and J. W. Tukey, "Modern Techniques of Power Spectrum Estimation," *IEEE Trans. Audio Electroacoustics*, AU-15, No. 2 (June 1967).

32. L. D. Enochson and A. G. Piersol, "Application of Fast Fourier Transform Procedures to Shock and Vibration Data Analysis," SAE Paper No. 670874, SAE, New York, 1967.

33. L. D. Enochson and R. K. Otnes, *Programming and Analysis for Digital Time Series Data*, Shock and Vibration Monograph Series, SVM-3, The Shock and Vibration Information Center, Dept. of Defense, Washington, D.C., 1968.

34. G. J. O'Hara, "A Numerical Procedure for Shock and Fourier Analysis," U.S. Naval Research Laboratory Report 5772, June 1962.

35. D. W. Lane, "Digital Shock Spectrum Analysis by Recursive Filtering," *Shock and Vibration Bull.* **33**, Part 2, 173–181, Shock and Vibration Information Center, 1964.

36. R. K. Otnes, "Single Degree of Freedom Filtering," MAC Report 500–02, Measurement Analysis Corporation, Los Angeles, Calif. (Sept. 1965).

37. D. J. Maglieri, "Some Effects of Airplane Operations and the Atmosphere on Sonic Boom Signatures," from NASA SP–147, *Sonic Boom Research*, 1967.

38. M. W. Oleson, "Integration and Double Integration – A Practical Technique," *Shock and Vibration Bull.* **35**, Part 4, 1–9, Shock and Vibration Information Center, 1966.

39. L. S. Jacobsen and R. S. Ayre, *Engineering Vibrations*, McGraw-Hill Book Co., New York, 1958.

40. R. S. Ayre, "Transient Response to Step and Pulse Functions," Vol. 1, Ch. 8, *The Shock and Vibration Handbook* (C. M. Harris and C. E. Crede, Eds.), McGraw-Hill Book Co., New York, 1961.

41. T. K. Hasselman, "An Analytical Basis For Time-Modulated Random Vibration Testing," NASA CR–66770 (Jan. 1969).

42. R. L. Barnoski and J. R. Maurer, "Mean Square Response of Simple Mechanical Systems to Nonstationary Random Excitation," *Trans. J. Appl. Mechanics*, Paper No. 69–APM–25, 1969.

43. C. G. Page, "Instantaneous Power Spectra," *J. Appl. Phys.* **23**, No. 1, 103–106 (Jan. 1952).

44. R. D. Kelly, "A Method for the Analysis of Short Duration Nonstationary Random Vibration," *Shock and Vibration Bull.* **29**, Part 4, 126–137, Dept. of Defense, 1961.

45. T. K. Caughey and H. J. Stumf, "Transient Response of a Dynamic System Under Random Loading, *J. Appl. Mechanics*, **28**, 565–566, Dec. 1961.

46. D. M. Aspinwall, "An Approximate Distribution for Maximum Response of a Random Vibration," Lockheed Missile and Space Co. Report No. TM 53–16 MD–5 (May 1961).

47. R. L. Barnoski and R. H. MacNeal, "The Peak Response of Simple Mechanical Systems to Random Excitation," CEA Project No. ES 182–6, technical report prepared by CEA for Lockheed Missile and Space Co. (Mar. 1962).

48. R. L. Barnoski, "The Maximum Response of a Linear Mechanical Oscillator to Stationary and Nonstationary Random Excitation," NASA CR–340 (Dec. 1965).

49. R. L. Barnoski, "Probabilistic Shock Spectra," to be published as a NASA CR (prepared under Contract NAS 1–8538 for NASA LRC, Dec. 1968).

50. A. H. Quazi, "Transient Signal Analysis by Means of Orthonormal

Functions," USNUSL Report No. 798, U.S. Navy Underwater Sound Laboratory, New London, Conn. (Feb. 1967).

51. F. B. Hildebrand, *Introduction to Numerical Analysis*, McGraw-Hill Book Co., New York, 1956.

52. C. E. Crede, "The Effect of Pulse Shape on Simple Systems Under Impulsive Loading," *Trans. ASME*, 77, 957–961 (Aug. 1955).

53. G. S. Mustin, *Theory and Practice of Cushion Design*, SVM–2, The Shock and Vibration Information Center, U.S. Department of Defense, Washington, D.C., 1968.

54. D. L. Cronin, "Application of Spectral Analysis to Spacecraft Design and Test," Report No. EM–17–19, TRW Systems, Redondo Beach, Calif. (Oct. 1967).

55. M. A. Biot and R. L. Bisplinghoff, "Dynamic Loads in Airplane Structures During Landing," NACA Report W–92 (ARR 4H10) (Oct. 1944).

56. Y. C. Fung and M. V. Barton, "Some Shock Spectra Characteristics and Uses," *J. Appl. Mechanics*, 25, 365–372 (Sept. 1958).

57. C. E. Crede, M. Gertel, and R. D. Cavanaugh, "Establishing Vibration and Shock Tests for Airborne Electronic Equipment," WADC Technical Report 54–272 (June 1954).

58. E. H. Schell, "Proximity Spectrum—A New Means of Evaluating Shock Motions," *Shock and Vibration Bull.* 35, Part 6, 229, Shock and Vibration Information Center, 1966.

59. E. H. Schell, "Applications of the Proximity Spectrum Including Spectra of Some Simple Shock Design and Test Motions," Air Force Flight Dynamics Report, AFFDL–TR–664, Wright-Patterson AFB, Ohio (May 1966).

60. Y. C. Fung, M. V. Barton, and D. Young, "Response of Nonlinear System to Shock Excitations," SAE Paper 585C, National Aerospace Engineering and Manufacturing Meeting, Los Angeles, Oct. 8–12, 1962.

61. T. F. Bogart, Jr., "Analog Method for Study of Shock Spectra in Nonlinear Systems," *Shock and Vibration Bull.* 35, Part 6, 197, Shock and Vibration Information Center, 1966.

62. W. B. Murfin, "Use of the Shock Spectrum Technique for Nonlinear Systems," *Shock and Vibration Bull.* 35, Part 7, 197–203, Shock and Vibration Information Center, 1966.

SUBJECT AND AUTHOR INDEX

Please note that the author entries appear in italics. The first number (in brackets) following the entry is the reference number. The second number is the page on which the reference is first cited.